AF776419

Imidazoline Inhibitors for Corrosion Protection of Oil Pipeline Steels

Imidazoline Inhibitors for Corrosion Protection of Oil Pipeline Steels

Experimental Laboratory Evaluation and Case Studies

Faysal Fayez Eliyan
Community College of Qatar
Doha, Qatar

WILEY

Copyright © 2024 by John Wiley & Sons, Inc. All rights reserved.

Published by John Wiley & Sons, Inc., Hoboken, New Jersey.
Published simultaneously in Canada.

No part of this publication may be reproduced, stored in a retrieval system, or transmitted in any form or by any means, electronic, mechanical, photocopying, recording, scanning, or otherwise, except as permitted under Section 107 or 108 of the 1976 United States Copyright Act, without either the prior written permission of the Publisher, or authorization through payment of the appropriate per-copy fee to the Copyright Clearance Center, Inc., 222 Rosewood Drive, Danvers, MA 01923, (978) 750-8400, fax (978) 750-4470, or on the web at www.copyright.com. Requests to the Publisher for permission should be addressed to the Permissions Department, John Wiley & Sons, Inc., 111 River Street, Hoboken, NJ 07030, (201) 748-6011, fax (201) 748-6008, or online at http://www.wiley.com/go/permission.

Trademarks: Wiley and the Wiley logo are trademarks or registered trademarks of John Wiley & Sons, Inc. and/or its affiliates in the United States and other countries and may not be used without written permission. All other trademarks are the property of their respective owners. John Wiley & Sons, Inc. is not associated with any product or vendor mentioned in this book.

Limit of Liability/Disclaimer of Warranty: While the publisher and author have used their best efforts in preparing this book, they make no representations or warranties with respect to the accuracy or completeness of the contents of this book and specifically disclaim any implied warranties of merchantability or fitness for a particular purpose. No warranty may be created or extended by sales representatives or written sales materials. The advice and strategies contained herein may not be suitable for your situation. You should consult with a professional where appropriate. Further, readers should be aware that websites listed in this work may have changed or disappeared between when this work was written and when it is read. Neither the publisher nor authors shall be liable for any loss of profit or any other commercial damages, including but not limited to special, incidental, consequential, or other damages.

For general information on our other products and services or for technical support, please contact our Customer Care Department within the United States at (800) 762-2974, outside the United States at (317) 572-3993 or fax (317) 572-4002.

Wiley also publishes its books in a variety of electronic formats. Some content that appears in print may not be available in electronic formats. For more information about Wiley products, visit our web site at www.wiley.com.

Library of Congress Cataloging-in-Publication Data

Names: Eliyan, Faysal Fayez, author.
Title: Imidazoline Inhibitors for Corrosion Protection of Oil Pipeline Steels : Experimental Laboratory Evaluation and Case Studies evaluation / Faysal Fayez Eliyan.
Description: Hoboken, New Jersey : Wiley, [2024] | Includes index.
Identifiers: LCCN 2023053482 (print) | LCCN 2023053483 (ebook) | ISBN 9781119437536 (cloth) | ISBN 9781119437550 (adobe pdf) | ISBN 9781119437567 (epub)
Subjects: LCSH: Petroleum pipelines–Corrosion. | Corrosion and anti-corrosives.
Classification: LCC TN879.55 .E55 2024 (print) | LCC TN879.55 (ebook) | DDC 620.1/1223–dc23/eng/20240109
LC record available at https://lccn.loc.gov/2023053482
LC ebook record available at https://lccn.loc.gov/2023053483

Cover Design: Wiley
Cover Image: © Matthias Kulka/Getty Images

Set in 9.5/12.5pt STIXTwoText by Straive, Pondicherry, India

To my loving family,
For being my unwavering support,
My constant source of inspiration,
This book is dedicated to you all.
Thank you for believing in me.

Contents

Preface

In the realm of corrosion protection, the need to safeguard oil pipelines against the relentless forces of degradation cannot be overstated. The economic and environmental consequences of pipeline failures demand the utmost attention and vigilance in implementing effective inhibition strategies. Among the myriads of corrosion inhibitors, imidazoline compounds have emerged as promising candidates for mitigating steel corrosion. However, in order to ascertain their true potential and optimize their performance, rigorous laboratory experimental evaluation is essential.

This book represents a comprehensive compilation of the experimental findings that delve into the evaluation of imidazoline inhibitors for safeguarding pipeline steels against corrosion. It serves as a definitive reference, consolidating invaluable insights and discoveries from a multitude of investigations. The experimental methodologies employed encompass a diverse range of techniques, enabling a thorough exploration of the inhibitive properties of imidazoline compounds.

The evaluation process includes an array of electrochemical measurements that provide crucial information on the corrosion behavior of pipeline steels. These measurements encompass potentiodynamic polarization, electrochemical impedance spectroscopy, open-circuit potentials, linear polarization, cyclic voltammetry, and electrochemical noise. By harnessing the power of these techniques, researchers have examined the intricate dynamics at play, shedding light on the efficacy of imidazoline inhibitors.

Additionally, the experimental evaluation delves into the microscopic realm, utilizing state-of-the-art tools to unravel the underlying mechanisms of corrosion inhibition. Techniques such as scanning electron microscopy (SEM) and profilometry, transmission electron microscopy (TEM), scanning Kelvin probe force microscopy (SKPFM), and scanning electrochemical microscopy (SECM) have enabled researchers to visualize and characterize the surfaces with unprecedented detail, thereby providing critical insights into the protective effects of imidazoline inhibitors.

Furthermore, spectroscopic tools have played a pivotal role in unraveling the molecular-level interactions between imidazoline inhibitors and pipeline steels. Techniques such as X-ray photoelectron spectroscopy (XPS), wire beam electrode (WBE), atomic force microscopy (AFM) characterization, scanning tunnel microscope, polarization modulation infrared reflection-absorption spectroscopy (PM-IRRAS), energy-dispersive X-ray spectroscopy (EDX), X-ray diffraction (XRD), and Fourier-transform infrared spectroscopy (FTIR) have been employed to elucidate the complex interplay between the inhibitors and the steel surfaces.

The book also delves into other experimental evaluations, including weight loss measurements and quartz crystal measurements, providing a comprehensive assessment of the corrosion inhibition effectiveness. Furthermore, computational studies, such as molecular dynamic simulations and Monte Carlo simulations, density functional theory, radial distribution function (RDF), and mean-square displacement (MSD), have been employed to augment the understanding of imidazoline inhibitors' performance.

Moreover, the book discusses the influence of physicochemical and flow effects, as well as metallurgical and surface effects, on the efficacy of imidazoline inhibitors. These effects encompass a broad range of factors, including chloride concentration, temperature, flow rate, acetic acid, and H_2S exposure. By exploring these complex variables, researchers have been able to elucidate the intricate interplay between corrosion inhibition and the environmental and operational conditions.

This book presents a collection of comprehensive case studies focusing on the corrosion control challenges faced by oil pipeline companies in operations. These case studies incorporate the details of managing pipelines with high gas content, considering factors such as temperature, flow regime, water content, and steel type. The book explores the significance of various corrosion control strategies, including the utilization of a variety of inhibitors, the implementation of pigging techniques, the application of cathodic protection, and the relevant codes and standards. Each case study examines realistic parameters, operation conditions, and provides detailed procedures for protecting, monitoring, and evaluating the performance of corrosion control measures. These cases serve as valuable references for industry professionals, offering insights into effective corrosion management in challenging oil pipeline environments.

In summary, this book provides a comprehensive and consolidated resource for understanding the experimental evaluation of imidazoline inhibitors on oil pipeline steels. It serves as a vital reference for corrosion scientists, engineers, and researchers, offering a wealth of knowledge and insights to enhance the corrosion protection strategies employed in the oil and gas industry. The invaluable findings contained within these pages will undoubtedly contribute to the ongoing pursuit of more robust and efficient corrosion inhibition solutions, ensuring the integrity and longevity of oil pipelines for years to come.

List of Abbreviations

2UEI	2-Undecyl-1-ethylamino imidazoline
AAOA	*N*-[2-[(2-Aminoethyl) amino] ethyl]-9-octadecenamide
ACDPAM	Water-soluble polyacrylamide modified by b-cyclodextrin
AEI-11	2-Undecyl-1-aminoethyl imidazoline
AFM	Atomic force microscopy
Ag	Silver
AgCl	Silver chloride
AM/DMDAAC/allyl-b-CD	Water-soluble polymer acrylamide/dimethyl diallyl ammonium chloride/allyl-b-CD
AM1	Austin Model 1
AP	Aqueous phase
API	American Petroleum Institute
AQI-11	2-Undecyl-1-aminoethyl-1-hydroxyethyl quaternary imidazoline
ASTM	American Society for Testing and Materials
AT-1	1,3-Bis(4-methoxybenzyl)-2-(4-methoxyphenyl)imidazoline
AT-2	1,3-Bis (4-methoxybenzyl)-2-phenylimidazolidine
AT-3	1,3-Bis(4-methoxybenzyl)-2-(4-nitrophenyl) imidazoline
BOP	Becke One Parameter
BTBBI	1-Butyl-2-thiobenzylbenzimidazole
c	Surface tension
$C_{18}H_{34}O_2$	Oleic acid
$C_3H_{10}N_2$	Trimethylenediamine
$C_4H_{10}N_2$	Butane-1,4-diamine
$C_4H_{12}N_2(NO_3)_2$	1,4-Diaminobutane dinitrate.

$C_4H_{12}N_2CO_3$	1,4-Diaminobutane carbonate
$Ca(OH)_2$	Calcium hydroxide
C_{dl}	Double-layer capacitance
CDs	Cyclodextrins
CFD	Computational fluid dynamics
CH_2	Methylene
CH_3	Methyl group
CH_4	Methane
Cl^-	Chloride ion
CMC	Critical micelle concentration
CO_2	Carbon dioxide
CO_3^{2-}	Carbonate
cP	Centipoise
CPE	Constant phase element
CPs	Chromeno pyrimidines
CR	Corrosion rate
cSt	Centistokes
CV	Cyclic voltammetry
D	Diffusion coefficients
DC	Direct current
DETA	Diethylenetriamine
DFT	Density functional theory
DBSI	Dodecylbenzene sulfonic acid imidazoline
DNP	Double Numeric plus Polarization
E_{corr}	Corrosion potential
EDX	Energy-dispersive X-ray spectroscopy
E_{HOMO}	Energy of highest occupied molecular orbital
EIS	Electrochemical impedance spectroscopy
E_{LUMO}	Energy of lowest unoccupied molecular orbital
EOI	Ethoxylated imidazoline
EN	Electrochemical noise
EWF	Electron work function
Fe	Metallic iron
$Fe(OH)_2$	Iron hydroxide
$Fe_2(SO_4)_3$	Mikasaite
Fe^{2+}	Ferrous ions
Fe^{3+}	Ferric ions
Fe_3C	Iron carbide
$FeCO_3$	Iron carbonate
FeOOH	Iron(III) oxide-hydroxide
FT-IR	Fourier-transform infrared spectroscopy
H	Hurst exponent

H_2CO_3	Carbonic acid
$H_2N\text{-}(CH_2)_4\text{-}NH_2$	1,4-Butanediamine
$H_2N\text{-}(CH_2\text{-}CH_2\text{-}NH)_4H$	Tetraethylene-pentamine
$H_2N\text{-}CH_2\text{-}CH_2\text{-}NH_2$	Ethylenediamine
H_2O	Water
H_2S	Hydrogen sulfide
H_2SO_4	Sulfuric acid
$H_3C\text{-}(CH_2)_{10}\text{-}COOH$	Lauric acid
$H_3C\text{-}(CH_2)_{12}\text{-}COOH$	Myristic acid
$H_3C\text{-}(CH_2)_{14}\text{-}COOH$	Palmitic acid
$H_3C\text{-}(CH_2)_{16}\text{-}COOH$	Stearic acid
H_3O^+	Hydronium ion
HAc	Acetic acid
HASI	1-[*N*,*N*′-bis(hydroxylethylether)-aminoethyl]-2-stearicimidazoline
HCO_3^-	Bicarbonate
HEI	Hydroxyethyl imidazoline
HNO_3	Nitric acid
i_{corr}	Corrosion current density
IM	Imidazoline
IM, CAS: 3010-23-9	1-(2-Aminoethyl)-2-heptadecyl imidazoline
IR	Infrared
KCl	Potassium chloride
LA	Lauric acid
LAI	Laurylamine imidazoline
LI	Localization index
LPR	Linear polarization resistance
MBT	Mercaptobenzothiazole
MEP	Molecular electrostatic potential
MSD	Mean-square displacement
N_2	Nitrogen
NIST REFPROP	NIST Reference Fluid Thermodynamic and Transport Properties
NaCl	Sodium chloride
$NaNO_2$	Sodium nitrite
NaOH	Sodium hydroxide
NMR	Nuclear magnetic resonance
NO	Nitric oxide
NO_2	Nitrogen dioxide
NOx	Nitrogen oxides
O_2	Oxygen
OCP	Open-circuit potential
OIM	Oleic-based imidazoline

ODA	Octadecylamine imidazoline
PAI	Palmitoylamine imidazoline
PAHs	Polycyclic aromatic hydrocarbons
PAM	Polyacrylamide
PGS-10	*N*-(3-((2*E*,4*E*)-5-(benzo[d][1,3]dioxol-5-yl) penta-2,4dienamido)propyl)-*N*,*N*-dimethyldecan-1-aminium iodide
PGS-12	*N*-(3-((2E,4E)-5-(benzo[d][1,3]dioxol-5-yl) penta-2,4-dienamido)propyl)-*N*,*N*-dimethyldodecan-1-aminium iodide
PGS-8	*N*-(3-((2*E*,4*E*)-5-(benzo[d][1,3]dioxol-5-yl) penta-2,4-dienamido) propyl)-*N*,*N*-dimethyloctan-1-aminium iodide
pH	Potential of hydrogen
Piperazine	C4H10N2
PM-IRRAS	Polarization modulation infrared reflection-absorption spectroscopy
ppm	Part per million
ppmv	Part per million by volume
PQI	1-(2-Aminoethyl)-1-dodecyl-2-hexadecyl-4,5-dihydro-1H-imidazol-1-ium
Pt/Ir	Platinum/iridium
PTFE	Polytetrafluoroethylene
PZC	Potential of zero charge
QSI	2-Heptadecyl-1-[2-(octadecanoylamino) ethyl]-2- imidazoline
Quaternary ammonium group	N+
R1	Long hydrocarbon chain (CH2)n–CH3 $n = 10–17$
R2	Active functional group (CH2)2–NH2)
RAIM	Rosin amide imidazoline
R_{ct}	Charge transfer resistance
RDF	Radial distribution function
R_n	Noise resistance
R_p	Polarization resistance
R_s	Solution resistance
R_t	Charge transfer resistance
SAI	Stearoylamine imidazoline
SB	Sodium benzoate
SC	Super critical
SCE	Standard calomel electrode
SECM	Scanning electrochemical microscopy
SEM	Scanning electron microscopy

SHE	Standard hydrogen electrode
SKPFM	Scanning Kelvin probe force microscopy
SO_2	Sulfur dioxide
SP	Super critical phase
SQI	1-(2-Aminoethyl)-1-dodecyl-2-heptadecyl-4,5-dihydro-1H-imidazol-1-ium
SSC	Stress corrosion cracking
STM	Scanning tunneling microscopy
TAI	1-(2-Thioureidoethyl)-2-alkylimidazoline
TAPS	Trans-Alaska Pipeline System
TEM	Transmission electron microscopy
TGA	Thioglycolic acid
TOFA	Tall oil fatty acid
UK	United Kingdom
VOCs	Volatile organic compounds
W	General corrosion rate
WBE	Wire beam electrode
WE	Working electrode
XPS	X-Ray photoelectron spectroscopy
XRD	X-Ray diffraction
ZRA	Zero resistance ammeter

List of Symbols

$\langle \rho_B \rangle_{local}$	Particle of *B* density averaged over all shells around particle *A*
ϕ_{sample}	Electron work function of sample
ϕ_{tip}	Electron work function of the electrode tip
ΔE	Energy gap
Δf	Resonant frequency shift
ΔG^{O}_{ads}	Standard free energy of adsorption
Δm	Change in mass per unit area
ΔN	Electron donation ability at the surface
μ	Dynamic viscosity
μ_d	Dipole moment
a	Frumkin interaction parameter
A	Electron affinity
API	Aminopropylimidazol
C_i	Inhibitor concentration
D	Differential polarization
E_a	Activation energy
$E_{inhibitor}$	Energy of inhibitor molecules
$E_{interaction}$	Energy of adsorption
$E_{surface+solution}$	Energy of interface
E_{total}	Inhibitor system energy
f	Temkin heterogeneity factor
f^-	Negative extent of Fukui functions
f^+	Positive extent of Fukui functions
f_{max}	Characteristic/maximum frequency
f_o	Fundamental frequency of the resonator
h	Average corrosion depth

I	Ionization potential
i_{corr}	Inhibited corrosion current density
i_{corro}	Uninhibited corrosion current density
K	Adsorption–desorption equilibrium constant
K_{ads}	Equilibrium constant of the adsorption reaction
$Org_{(sol)}$	Organic inhibitors in the aqueous phase
$q_{(N)}$	Mulliken atomic charges of the atom in molecules with N electrons
$q_{(N+1)}$	Mulliken atomic charges of the atom in molecules with $N+1$ electrons
$q_{(N-1)}$	Mulliken atomic charges of the atom in molecules with $N-1$ electrons
Q_{ads}	Heat of adsorption
R	Universal gas constant
Re_c	Local Reynolds number
T	Temperature
t	Immersion time
u_c	Local velocity
V	Potential difference
V_{corr}	Localized corrosion rate
W	General corrosion rate
Z_w	Warburg resistance
β	Polarization slope
ΔEL-H	Energy difference
ΔN	Number of transferred electrons from an inhibitor molecule to the carbon steel
η	Inhibition efficiency
η_G	Global hardness
η_{GFe}	Global hardness of iron
η_{inh}	Hardness of the inhibitor molecule
θ	Coverage of corrosion inhibitor of the surface
μ_q	Shear modulus of the quartz crystal
ν	Kinematic viscosity
ρ	Density of test solution
ρ_q	Density of the quartz crystal
σ	Resistance of diffusion through inhibitor film
σ_i	Current noise standard deviation
σ_s	Softness
σ_v	Potential noise standard deviation

τ_c Shear stress
χ Electronegativity
χ_{Fe} Electronegativity of iron
χ_{inh} Electronegativity of synthesized inhibitor molecule
ω Rotational speed
ω_o Warburg coefficient

1

Introduction

1.1 Importance of Corrosion Inhibition in Pipeline Systems

The importance of corrosion inhibition in pipeline systems cannot be overstated, as corrosion poses significant challenges to the integrity, reliability, and safety of these critical infrastructure components. Pipelines are vital for the transportation of oil, gas, and other fluids over long distances, often spanning diverse and harsh environments. The exposure of pipeline steels to corrosive conditions, such as carbon dioxide (CO_2), hydrogen sulfide (H_2S), oxygen, and various contaminants, makes them susceptible to corrosion, which can lead to severe consequences. Corrosion in pipelines can result in significant economic losses. The oil and gas industry heavily relies on efficient and reliable pipeline networks to transport valuable resources from production sites to refineries and end-users. Corrosion-induced failures can lead to costly downtime, production losses, and repair or replacement expenses. The financial impact extends beyond direct costs, encompassing environmental remediation, legal liabilities, and damage to a company's reputation. Therefore, effective corrosion inhibition strategies are essential to minimize these financial burdens and maintain the long-term operability of pipeline systems. Moreover, corrosion inhibition plays a crucial role in ensuring the safety of pipeline operations. Corroded pipelines may experience structural weakening, leading to leaks, ruptures, or catastrophic failures. These failures not only endanger personnel working in the vicinity but also pose significant risks to nearby communities and the environment. The release of flammable or toxic substances can result in fires, explosions, or environmental pollution. By employing effective corrosion inhibitors, the overall structural integrity of pipeline systems can be

Imidazoline Inhibitors for Corrosion Protection of Oil Pipeline Steels: Experimental Laboratory Evaluation and Case Studies, First Edition. Faysal Fayez Eliyan.
© 2024 John Wiley & Sons, Inc. Published 2024 by John Wiley & Sons, Inc.

preserved, mitigating the potential for such hazardous incidents and safeguarding human lives.

Corrosion inhibition is also vital for maintaining the efficiency and reliability of pipeline networks. Corroded pipelines exhibit increased frictional resistance, causing flow restrictions, pressure drops, and decreased operational efficiency. Additionally, the formation of corrosion by-products, such as scales and deposits, can further impede fluid flow and reduce pipeline capacity. By employing corrosion inhibitors, the formation and accumulation of these deleterious products can be mitigated, ensuring smooth and efficient flow throughout the pipeline network. This, in turn, optimizes operational performance, reduces energy consumption, and enhances the overall reliability of the transportation system. Furthermore, effective corrosion inhibition strategies contribute to environmental protection. Corroded pipelines can lead to leaks or spills, resulting in the release of hydrocarbons or other hazardous substances into the surrounding environment. These incidents can have severe ecological impacts, contaminating soil, water bodies, and sensitive ecosystems. By employing corrosion inhibitors that effectively mitigate corrosion, the risk of environmental pollution can be significantly reduced, preserving natural resources and minimizing the ecological footprint of the oil and gas industry.

Several oil pipeline incidents have occurred due to problems with corrosion inhibition. In March 2006, the Trans-Alaska Pipeline System (TAPS) experienced a leak caused by a failure in the corrosion inhibition system, resulting in the spillage of approximately 200,000 gallons of crude oil near Atigun Pass. In July 2010, the Lakehead Pipeline System operated by Enbridge Energy Partners ruptured in Michigan due to corrosion issues stemming from inadequate corrosion inhibition, releasing an estimated one million gallons of heavy crude oil into the Kalamazoo River. The Platte Pipeline, owned by Kinder Morgan Energy Partners, suffered a release in April 2011 near Atwood, Kansas, also attributed to corrosion problems arising from inadequate corrosion inhibition methods, resulting in the spillage of around 42,000 gallons of crude oil. Around the same time, in April 2011, the Rainbow Pipeline owned by Plains Midstream Canada experienced a leak in northern Alberta, Canada, due to corrosion issues, leading to the release of an estimated 28,000 barrels (1.18 million gallons) of crude oil, causing contamination and harm to the environment. These incidents prompted cleanup operations, repairs, and investigations into the corrosion inhibition measures employed by the respective pipeline operators. Please note that this information is based on historical data up until September 2021, and there may have been additional incidents since then.

1.2 The Economic Consequences of Corrosion in the Oil and Gas Industry

1.2.1 Maintenance Costs

Maintenance costs associated with corrosion in the oil and gas industry can impose substantial financial burdens on companies. Corrosion is an ongoing challenge that requires regular inspection, maintenance, and repair of pipelines and equipment to mitigate its detrimental effects. The continuous monitoring and evaluation of corrosion conditions necessitate dedicated resources, including specialized personnel, equipment, and technologies. These costs encompass the procurement and operation of inspection tools, such as sensors, cathodic protection systems, and non-destructive testing equipment. Inspection activities are critical for identifying corrosion-prone areas, assessing the extent of corrosion damage, and determining the need for repair or replacement. Regular inspections often involve detailed examinations of pipeline surfaces, internal inspections using intelligent pigging techniques, and assessments of corrosion inhibitors' effectiveness. These inspection processes require skilled personnel, sophisticated data analysis, and comprehensive reporting, all of which contribute to the overall maintenance costs. Furthermore, maintenance and repair efforts entail significant expenses. Corrosion-related repairs can range from simple patching or coating repairs to more complex tasks like section replacements or full pipeline rehabilitation. The costs involved in sourcing materials, specialized coatings, and corrosion inhibitors, as well as the labor-intensive nature of repair activities, contribute to the financial burden. Additionally, access to remote or offshore locations may require specialized transportation and logistical arrangements, further adding to the overall maintenance costs. Corrosion-induced maintenance expenses are not limited to pipelines alone. Equipment, such as storage tanks, valves, pumps, and processing units, are also susceptible to corrosion and require regular inspection and upkeep. Failure to address corrosion in these critical components can lead to significant disruptions in operations, safety risks, and potential environmental incidents. Therefore, oil and gas companies allocate substantial financial resources to ensure proper maintenance, including equipment inspection, cleaning, repair, and replacement as necessary.

1.2.2 Downtime and Production Losses

Downtime and production losses are major consequences of corrosion in the oil and gas industry. When pipelines require repairs or replacements due to corrosion-related damage, it leads to interruptions in production and transportation, resulting in reduced output and potential revenue losses for the companies involved. These disruptions can have far-reaching impacts on the entire supply chain and the overall profitability of oil and gas operations. The process of repairing or replacing corroded pipelines is time-consuming and complex. It involves identifying the affected sections, assessing the extent of damage, procuring materials, planning the repair activities, and executing the necessary repairs. Depending on the severity of the corrosion, these repairs can range from localized patching to complete pipeline segment replacements. This downtime can last for days, weeks, or even months, during which the flow of oil and gas is halted or significantly reduced. The interruption of production and transportation due to corrosion-related repairs has ripple effects throughout the industry. Oil and gas companies rely on a continuous flow of resources to meet production targets and supply customer demands. When pipelines are out of commission, there is a disruption in the movement of hydrocarbons from production fields to processing facilities, refineries, and ultimately to end consumers. This disruption can lead to delays in meeting contractual obligations, potential penalties, and strained relationships with customers. Moreover, the reduction in output directly impacts revenue generation. Oil and gas companies earn revenue based on the quantity of oil and gas sold. When production is curtailed due to corrosion-related issues, the volume of hydrocarbons available for sale decreases, leading to revenue losses. Additionally, companies may face increased costs associated with rerouting or alternative transportation methods to compensate for the interrupted pipelines, further impacting their profitability. The economic consequences of downtime and production losses extend beyond the immediate revenue impact. They can also affect market competitiveness and investor confidence. If a company experiences frequent disruptions due to corrosion-related repairs, its reputation and reliability in the industry may suffer. This could lead to decreased investor interest, difficulty in securing future projects, and potential negative impacts on stock prices.

To mitigate the impact of downtime and production losses, oil and gas companies invest in proactive corrosion management strategies. This includes implementing corrosion monitoring systems, employing effective corrosion inhibitors, conducting regular inspections, and adopting preventive maintenance practices. By prioritizing corrosion prevention and adopting proactive measures, companies can minimize the frequency and severity of corrosion-related issues, thereby reducing downtime, optimizing production levels, and safeguarding their bottom line.

1.2.3 Replacement Expenses

Corrosion-related damage in the oil and gas industry can have substantial financial implications, particularly when it requires the complete replacement of pipelines, storage tanks, or other critical infrastructure components. The decision to undertake replacements is often driven by the severity of corrosion, which can compromise the safety, reliability, and efficiency of these assets. As a result, companies are faced with significant expenses associated with replacing corroded equipment to ensure the continuity of their operations. The process of replacing corroded pipelines or storage tanks involves multiple stages, each contributing to the overall expenses. First, thorough inspections and assessments are conducted to evaluate the extent of corrosion damage and identify the areas that require replacement. Once the areas requiring replacement are identified, companies must procure the necessary materials and equipment. This entails sourcing new pipes, fittings, valves, or storage tanks, often with specific requirements to address corrosion concerns. The costs of these materials can be substantial, especially for large-scale projects that involve significant lengths of pipelines or the replacement of multiple storage tanks. Additionally, specialized coatings or corrosion-resistant alloys may be required to ensure the longevity and durability of the new infrastructure.

1.2.4 Increased Energy Consumption

Corrosion in the oil and gas industry not only leads to direct material damage and replacement expenses but also has indirect consequences, such as increased energy consumption. When pipelines and equipment are affected by corrosion, it can result in reduced flow efficiency and increased resistance to fluid movement. As a result, more energy is required to overcome these flow restrictions and maintain optimal transportation of oil and gas through the corroded infrastructure. Corrosion-induced flow restrictions create additional pressure drop along the pipeline, necessitating higher pumping or compression power to maintain the desired flow rates. This increased energy consumption translates into higher operating costs for oil and gas companies. Inefficiencies caused by corrosion can require operators to increase pump or compressor discharge pressures to compensate for the decreased flow capacity, resulting in higher energy demands. Furthermore, if corrosion restricts the flow beyond the acceptable limits, companies may need to operate at reduced throughput, leading to extended transportation times and increased energy usage per unit of product transported. The overall impact of increased energy consumption due to corrosion extends beyond operational costs. It contributes to higher carbon emissions and environmental footprints. The additional energy required to compensate for corrosion-induced inefficiencies results in increased greenhouse gas emissions, primarily from the combustion of fossil fuels used for power generation. This contributes to the overall carbon intensity of the oil and gas industry.

1.3 The Environmental Consequences of Corrosion in the Oil and Gas Industry

1.3.1 Soil and Water Contamination

Soil and water contamination is a significant environmental consequence of corrosion in the oil and gas industry. Corrosion-related leaks, breaches, or ruptures in pipelines or storage tanks can result in the unintended release of oil or gas into the surrounding environment, posing risks to soil quality, groundwater, and nearby water bodies. This form of environmental pollution can have severe consequences for ecosystems, biodiversity, and human health. When corrosive damage causes leaks or ruptures, oil or gas substances can seep into the soil, leading to soil contamination. The spilled hydrocarbons can infiltrate the soil layers, altering its composition, reducing its fertility, and disrupting its natural processes. The toxic components present in the spilled oil or gas, such as heavy metals, volatile organic compounds (VOCs), or polycyclic aromatic hydrocarbons (PAHs), can persist in the soil for extended periods, posing risks to plant and microbial life. These contaminants can also leach into groundwater, further exacerbating the environmental impact. Water bodies, including rivers, lakes, and coastal areas, are particularly vulnerable to contamination from corrosion-related leaks. Spilled oil or gas can enter waterways through surface runoff or direct discharge, leading to immediate or long-term environmental damage. The oil or gas substances can form slicks or sheens on the water surface, affecting oxygen exchange and sunlight penetration, which are essential for aquatic life. The toxic components of the hydrocarbons can harm fish, shellfish, and other organisms: disrupt food chains, and disturb the ecological balance of the affected ecosystems.

Contaminated water sources pose risks to human health, especially when they are used for drinking, agriculture, or recreational purposes. Exposure to contaminated soil or water can lead to adverse health effects, including respiratory problems, skin irritations, or contamination of food crops. The economic impacts can also be substantial, as contaminated water bodies may result in restrictions on fishing or tourism activities, leading to revenue losses for local communities and businesses.

1.3.2 Air Pollution

Air pollution is one of the major consequences of corrosion in the oil and gas industry. Corrosion-related incidents, such as leaks, ruptures, or emissions from equipment, can release various pollutants into the atmosphere, contributing to air pollution. These pollutants include VOCs, sulfur compounds, nitrogen oxides (NOx), and particulate matter. The release of such contaminants can have detrimental effects on air quality, human health, and the environment. VOCs are organic chemicals that readily vaporize at room temperature and contribute to the formation of ground-level ozone and secondary pollutants. These compounds have harmful effects on air quality and human health, causing respiratory issues, eye irritation, and even long-term health risks such as cancer. The release of VOCs into the atmosphere also contributes to the formation of smog and photochemical reactions, leading to the creation of harmful air pollutants. Furthermore, corrosion-induced emissions can also result in the release of sulfur compounds, such as H_2S and NOx. These gases contribute to the formation of acid rain and atmospheric pollution. Sulfur compounds released from corroded infrastructure contribute to the generation of sulfur dioxide (SO_2) when they react with oxygen in the air. SO_2 is a major contributor to acid rain, which can have harmful effects on ecosystems, soil quality, and vegetation. Similarly, NOx emissions, primarily from combustion processes in equipment affected by corrosion, contribute to the formation of nitrogen dioxide (NO_2) and other NOx. The particulate matter released as a result of corrosion-related processes can also contribute to air pollution. Particulate matter consists of fine particles suspended in the air, including dust, soot, and other solid or liquid pollutants. When corrosion causes equipment or pipelines to deteriorate, particulate matter can be released into the atmosphere, especially during maintenance activities or repairs. These particles can have adverse effects on air quality and human health, as they can be inhaled and deposited in the respiratory system, leading to respiratory problems, cardiovascular issues, and other health complications.

1.3.3 Climate Change Contributions

Corrosion in the oil and gas industry plays a role in contributing to climate change. The extraction, production, and transportation of oil and gas involve energy-intensive processes that release greenhouse gas emissions into the atmosphere. Corrosion-related incidents exacerbate this issue by causing leaks, spills, or inefficient operations, resulting in additional emissions of CO_2 and other greenhouse gases. These emissions contribute to the accumulation of greenhouse gases in the atmosphere, leading to the greenhouse effect and subsequent climate change. Corrosion-related leaks in pipelines or storage tanks release methane (CH_4), which is a potent greenhouse gas with a much higher global warming potential than CO_2 over shorter timeframes. Methane emissions occur due to the anaerobic decomposition of organic matter or the incomplete combustion of hydrocarbons. When corrosion compromises the integrity of pipelines or storage facilities, it can lead to the release of methane into the atmosphere, amplifying the climate change impact of the oil and gas industry. Inefficient operations resulting from corrosion also contribute to climate change. When equipment is corroded, it can lead to suboptimal combustion or energy utilization, resulting in higher energy consumption and associated greenhouse gas emissions. For example, corroded equipment may experience reduced thermal efficiency, requiring more fuel or energy input to achieve the desired heat transfer, which in turn increases CO_2 emissions. Similarly, corrosion-induced leaks in pressure vessels or valves can lead to fugitive emissions of methane or other greenhouse gases, further contributing to the industry's carbon footprint.

The impacts of climate change, such as rising global temperatures, sea level rise, and extreme weather events, have far-reaching consequences for ecosystems, communities, and economies. The oil and gas industry's contribution to climate change through corrosion-related emissions intensifies these risks. It underscores the importance of adopting corrosion prevention measures, implementing energy-efficient practices, and transitioning toward cleaner and renewable energy sources. By investing in technologies that reduce emissions, promoting sustainable practices, and supporting research and development efforts for low-carbon solutions, the industry can mitigate its climate change contributions and contribute to a more sustainable future.

1.4 Importance of Imidazoline Inhibitors in Oil Pipelines

Imidazoline inhibitors play a crucial role in maintaining the integrity and efficiency of oil pipelines. These chemical compounds are specifically designed to prevent the formation and accumulation of harmful deposits, such as scales, corrosion, and organic sludges, within the pipeline system. By inhibiting the formation of these deposits, imidazoline inhibitors help ensure the smooth flow of oil, minimize downtime, and enhance the overall safety and reliability of oil pipelines. One of the primary reasons imidazoline inhibitors are essential in oil pipelines is their ability to mitigate corrosion. Imidazoline inhibitors create a protective film on the internal surface of pipelines, which acts as a barrier between the metal and corrosive substances present in the oil. This protective film reduces the corrosion rate and extends the lifespan of the pipeline, resulting in significant cost savings for oil companies and reducing the risk of catastrophic failures. Another key aspect of imidazoline inhibitors is their capacity to inhibit the formation of scales. In oil production, scales can precipitate and accumulate on the inner walls of pipelines. This build-up restricts the flow of oil, reduces pipeline capacity, and increases pumping costs. Imidazoline inhibitors prevent the nucleation and growth of scale crystals, effectively controlling their deposition and maintaining the hydraulic efficiency of the pipeline. Imidazoline inhibitors are also effective in controlling the formation of organic sludges in oil pipelines. Sludges can form due to the presence of asphaltenes, waxes, and other heavy organic compounds in the oil. These sludges can cause blockages, increase pressure differentials, and hinder the flow of oil. Imidazoline inhibitors interact with organic components, dispersing and preventing their agglomeration, thereby minimizing sludge formation and maintaining the pipeline's operational efficiency. They ultimately aid in reducing the frequency of pipeline maintenance and cleaning operations. By inhibiting the deposition of scales, corrosion, and sludges, these inhibitors help prevent the accumulation of harmful substances that would otherwise require costly and time-consuming interventions. This leads to improved operational efficiency, reduced downtime, and enhanced productivity in oil pipeline operations.

Imidazoline inhibitors are utilized in oil pipelines through a process known as chemical injection. This involves the controlled addition of imidazoline-based compounds into the pipeline system at specific points, such as injection stations or at strategic intervals along the pipeline route. The inhibitors are typically introduced into the pipeline through injection skids or dosing units, which accurately measure and regulate the dosage of the inhibitor. Once injected into the pipeline, the imidazoline inhibitors mix with the flowing oil and disperse throughout the system. As the oil flows, the inhibitors come into contact with the internal surfaces of the pipeline, where they begin to perform their protective functions. Imidazoline inhibitors possess a unique molecular structure that allows them to adsorb onto the metal surfaces, forming a thin, protective film. This protective film acts as a barrier, preventing corrosive substances from directly interacting with the pipeline's metal surfaces. It hinders the electrochemical reactions that lead to corrosion, reducing the corrosion rate and mitigating the risk of metal degradation. Additionally, imidazoline inhibitors possess surfaceactive properties, which enable them to interact with and inhibit the formation of scales and organic sludges. The dosage of imidazoline inhibitors used in oil pipelines varies depending on factors such as the pipeline size, operating conditions, and the specific inhibitor formulation. The dosage is carefully determined through extensive laboratory testing and field trials to ensure optimal performance and effectiveness. Industry professionals consider factors such as the pipeline's material, fluid composition, flow rates, and environmental conditions when establishing the appropriate dosage. Regular monitoring and maintenance of the imidazoline inhibitor treatment program are essential to ensure its continued effectiveness. Periodic sampling and analysis of the treated oil help assess the inhibitor's performance, corrosion rates, and the presence of any undesirable deposits. Adjustments to the dosage or inhibitor formulation may be made based on the analysis results to optimize protection and maintain the pipeline's integrity.

1.5 Importance of Experimental Laboratory Evaluation of Imidazoline Inhibitors

Laboratory experimental evaluation plays a crucial role in assessing the effectiveness of imidazoline inhibitors in mitigating corrosion of pipeline steels. These evaluations provide essential insights on the following:

- Corrosion assessment: Laboratory experiments allow researchers to simulate and evaluate the corrosion behavior of pipeline steels in controlled environments. By subjecting the steel samples to corrosive conditions representative of the pipeline's operational environment, researchers can assess the extent of corrosion and evaluate the performance of imidazoline inhibitors. This helps in understanding the inhibitor's ability to mitigate corrosion, measure the corrosion rate reduction, and determine its long-term protective efficacy.
- Inhibitor formulation optimization: Laboratory evaluations enable researchers to optimize the formulation of imidazoline inhibitors. By testing different concentrations and combinations of inhibitors, researchers can identify the most effective formulation that provides maximum corrosion inhibition. These experiments also help determine the optimal dosage range for the inhibitors, ensuring cost-effective use while maintaining optimal corrosion protection for the pipeline steels.
- Material compatibility: Laboratory evaluations allow researchers to examine the compatibility of imidazoline inhibitors with different types of pipeline steels. It is important to ensure that the inhibitors do not adversely affect the mechanical properties or integrity of the steel. By

conducting tests such as tensile strength analysis, hardness testing, and microstructural examinations, researchers can assess any potential degradation or changes in the steel's properties due to the utilization of inhibitors.

- Performance in various environmental conditions: Laboratory experiments provide researchers with the flexibility to simulate various environmental conditions that pipelines may encounter. Different corrosive media, such as acidic or saline solutions, can be replicated to evaluate the performance of imidazoline inhibitors in different scenarios. This helps in understanding how the inhibitors behave under different conditions and allows for targeted inhibitor selection based on the specific pipeline's operating environment.
- Comparative studies: Laboratory evaluations facilitate comparative studies between different imidazoline inhibitors or inhibitor formulations. By testing multiple inhibitors simultaneously, researchers can determine which formulation offers superior corrosion protection for the pipeline steels. These comparative studies help in the selection of the most effective inhibitor and provide valuable information for industry professionals seeking to optimize corrosion mitigation strategies.

1.6 Experimental Considerations for Imidazoline Inhibitors in Oil Pipeline Applications

Before imidazoline inhibitors are considered for deployment, a thorough evaluation of their performance experimentally is imperative. This comprehensive guide explores the key factors that demand attention when assessing the suitability of imidazoline inhibitors for use in oil pipelines. They include performance evaluation and compatibility with pipeline fluids to environmental impact assessment, material compatibility, long-term performance, and field testing validation. By understanding and addressing these experimental considerations, pipeline operators can make informed decisions about the application of imidazoline inhibitors, ensuring effective corrosion and scale protection throughout their pipeline systems.

- Inhibitor performance evaluation: In a laboratory setting, it is essential to conduct comprehensive tests to evaluate the inhibitory performance of imidazoline compounds. These tests should assess their effectiveness in mitigating corrosion and scale formation. Simulating pipeline-relevant conditions, such as temperature, pressure, and fluid composition, will provide valuable insights into the inhibitors' performance under real-world circumstances.

- Compatibility with pipeline fluids: Imidazoline inhibitors must be compatible with the various types of fluids present in oil pipelines. Compatibility studies should be conducted to determine how the inhibitors interact with crude oil and its associated contaminants. It is important to assess the stability and efficacy of the inhibitors in the presence of impurities like salts, solids, and acidic components commonly found in pipeline fluids.
- Environmental considerations: The environmental impact of imidazoline inhibitors must be evaluated to ensure compliance with regulatory requirements. Toxicity studies should be conducted to assess the potential effects of these inhibitors on aquatic life and other environmental factors. Understanding and minimizing any adverse environmental impacts are critical aspects of their overall suitability for pipeline use.
- Material compatibility: Compatibility with pipeline materials, particularly steel, is of utmost importance. Investigating the potential interaction between imidazoline inhibitors and pipeline materials helps avoid any adverse effects or degradation. Corrosion testing on pipeline materials exposed to the inhibitors can provide insights into their long-term compatibility and ensure the structural integrity of the pipeline.
- Long-term performance and compatibility: Imidazoline inhibitors need to demonstrate reliable performance and stability over the long term. Conducting extended tests or simulating accelerated aging conditions allows for an evaluation of their durability and effectiveness over extended periods. Understanding any potential depletion or degradation of the inhibitors and developing strategies for replenishment or maintenance are essential for sustained protection of the pipeline.
- Cost-effectiveness and feasibility: Evaluating the cost-effectiveness and feasibility of implementing imidazoline inhibitors in oil pipelines is crucial. Considerations should include the production, transportation, and application requirements of the inhibitors. Understanding their availability, cost, and compatibility with existing pipeline infrastructure ensures an informed decision-making process.
- Field testing and validation: Field trials are necessary to validate the performance of imidazoline inhibitors under real-world pipeline conditions. Monitoring their effectiveness and evaluating any associated operational considerations during field testing provide practical insights. Field data helps verify the laboratory findings and allows for adjustments or optimizations if needed.
- Documentation and reporting: Thorough documentation of experimental procedures, test results, and observations is essential. Detailed reports

summarizing the findings, including recommendations for further optimization or considerations for full-scale implementation, aid in decision-making and knowledge sharing within the industry.

1.7 Procedure for Experimental Laboratory Evaluation of Imidazoline Inhibitors for Utilization in Oil Pipelines

The evaluation of performance in the laboratory is a crucial step in determining the suitability of imidazoline inhibitors for application in oil pipelines. Before considering their implementation in real-world scenarios, it is imperative to subject these inhibitors to rigorous laboratory testing to assess their effectiveness, compatibility, and potential risks. This evaluation process involves various tests and analyses on the inhibitor's performance, allowing engineers and researchers to make informed decisions regarding their application. Listed below is the procedure for evaluating the performance of imidazoline inhibitors in the laboratory, highlighting the significance of these assessments in ensuring the reliability and safety of oil pipelines.

- Experiment type:
  A controlled laboratory study shall be conducted to evaluate the effectiveness of imidazoline inhibitors in preventing, mitigating, and controlling corrosion in oil pipelines. Pipeline steel samples shall be exposed to various flow conditions, and the corrosion rates shall be analyzed with and without the imidazoline inhibitors. It shall be noted that laboratory conditions may not fully replicate the complexities and dynamic nature of real-world pipeline systems. Additionally, the results obtained from the laboratory study may have limitations in terms of generalizability to actual pipeline operations due to the controlled environment and potential differences in scale and operational conditions.
- Laboratory setups:
  i) The corrosion cell setup shall be prepared according to recognized standards such as ASTM G31 [NACE ASTM, 2012] or NACE TM0169 [NACE International, 2016]. The cell shall consist of a cylindrical vessel with provisions for introducing the oil, water, and inhibitor solution. The cell shall have an electrode system to measure the corrosion rates and other auxiliary setups such as a temperature control system. It is essential to ensure that the setup accurately represents the conditions and variables being studied. The laboratory setup may not perfectly mimic the actual pipeline environment, and certain factors

such as the absence of flowing oil and pipeline-specific configurations may limit the representativeness of the results obtained.

ii) Representative samples of the pipeline steel to be used in the experiment shall be obtained. The samples shall be cleaned and ensured to be free from any corrosion products or contaminants. The selection of samples can be guided by industry standards like API 5L [American Petroleum Institute, 2018] or ISO 3183 [International Organization for Standardization, 2019A]. The selection of representative pipeline steel samples is crucial to ensure the applicability of the study's findings. It should be noted that the pipeline steel samples used in the laboratory may not fully capture the variations and heterogeneity observed in real pipeline systems. The results shall be interpreted within the context of the specific steel samples used.

- Laboratory parameters:
  i) The appropriate statistical design for the experiment shall be determined, such as a randomized complete block design or a factorial design, depending on the factors and variables to be studied. Statistical guidelines such as ASTM E122 [American Society for Testing and Materials, 2009A] or ISO 5725 [International Organization for Standardization, 2019B] shall be consulted for experimental design and analysis. The statistical design chosen for the laboratory study may not account for all potential factors and interactions present in actual pipeline environments. The results shall be interpreted considering these limitations.
  ii) The number of pipeline steel samples required for each experimental condition shall be determined, considering the statistical power and significance level desired. The sample size shall be calculated using established methods like power analysis as per ASTM E1225 [American Society for Testing and Materials, 2019B] or ISO 5725 [International Organization for Standardization, 2019B]. The sample size in laboratory studies is often limited due to resource constraints and practical considerations. The limited sample size may impact the statistical power and precision of the study's findings.
  iii) The duration of exposure for each sample shall be decided to simulate real-life conditions and allow sufficient time for corrosion to occur. The exposure time can be determined based on industry guidelines such as ASTM G31 or NACE TM0169 [NACE International, 2016]. The exposure time in the laboratory may differ from the long-term exposure experienced in real pipeline operations. It is important to consider this difference when interpreting the results.

iv) The number of replicates for each experimental condition shall be decided to ensure statistical validity and reduce variability. Guidelines from ASTM E122 [American Society for Testing and Materials, 2009A] or ISO 5725 [International Organization for Standardization, 2019B] shall be considered for appropriate replication. The number of replicates in laboratory studies is often limited compared to the large number of data points available in real pipeline monitoring. The limited number of replicates may affect the precision and robustness of the study's findings.

- Statistical analysis:
  i) The corrosion rates of the pipeline steel samples for each experimental condition shall be recorded. Accurate and consistent data collection shall be ensured following recognized procedures such as ASTM G1 [American Society for Testing and Materials, 2019D] or NACE TM0169 [NACE International, 2016]. Laboratory measurements may not capture all the variations and complexities present in actual pipeline environments, which can introduce some limitations and uncertainties to the study's results.
  ii) Statistical analysis shall be performed using appropriate techniques such as analysis of variance (ANOVA) to determine the significance of the imidazoline inhibitors in corrosion prevention, mitigation, and control. The data shall be analyzed using statistical software packages like SPSS, Minitab, or R, following guidelines from ASTM E178 [American Society for Testing and Materials, 2019C]. The statistical analysis conducted in the laboratory study may not fully account for all the factors and interactions present in real-world pipeline systems. The results shall be interpreted while considering these limitations.
  iii) The results shall be analyzed to draw meaningful conclusions and determine the effectiveness of the imidazoline inhibitors. Confidence intervals, p-values, and effect sizes shall be considered while interpreting the data, following guidelines from ASTM E178 [American Society for Testing and Materials, 2019C]. The interpretation of laboratory results shall be done with caution, considering the limitations inherent in the laboratory setup and the potential differences from real-world pipeline conditions.
- Imidazoline inhibitor type:
  The specific type of imidazoline inhibitor to be evaluated shall be selected based on prior knowledge or a literature review. Commercially available inhibitors that comply with relevant industry standards, such as NACE Standard RP0775 [NACE International, 2017] or ISO 12944 [International

Organization for Standardization, 2017], shall be considered. The chosen imidazoline inhibitor may have specific characteristics or limitations that could impact its performance and generalizability to different pipeline systems. It is important to consider these limitations when interpreting the results.

- Pipeline steel type:
  Pipeline steel samples that are representative of the steel used in actual oil pipelines shall be chosen. Factors such as composition, microstructure, and mechanical properties shall be considered to ensure realistic and applicable results. Industry standards like API 5L [American Petroleum Institute, 2018] or ISO 3183 [International Organization for Standardization, 2019A] shall be referred to for guidance on pipeline steel specifications. The laboratory study may not account for the variations and specific characteristics of pipeline steel used in different real-world pipeline systems. The results shall be interpreted considering the specific steel samples used in the study.
- Flow rate:
  The desired flow rate shall be determined based on the operating conditions of oil pipelines under evaluation. The flow rate shall be representative of the typical flow rates encountered in practice. Industry guidelines shall be considered for determining flow rates. The laboratory study's chosen flow rate may not fully replicate the flow dynamics and variations observed in real-world pipeline operations. The results shall be interpreted within the context of the specific flow rate used in the study.
- Flow chemical composition:
  A synthetic oil solution with a chemical composition similar to that of the oil expected to flow through the pipelines shall be prepared. The composition shall include the relevant hydrocarbons, additives, impurities, and contaminants. Industry standards shall be referred to for guidance on oil composition and contaminants. The synthetic oil solution used in the laboratory study may not perfectly replicate the complex and variable chemical composition of real oil flows. The results shall be interpreted considering this limitation and potential variations in actual oil compositions.
- Flow regimes:
  Different flow regimes shall be simulated to mimic the conditions experienced in oil pipelines. These may include different flow types, depending on the expected flow patterns in the pipeline. Recognized guidelines shall be followed for determining flow regimes. The laboratory study may not fully replicate the dynamic and complex flow regimes observed in real oil

pipelines. The results shall be interpreted while considering the limitations in replicating all possible flow regimes.

- Flow dissolved gases:
  The presence of dissolved gases in the oil flow, such as oxygen, CO_2, and H_2S, shall be considered. These gases shall be incorporated into the synthetic oil solution to replicate realistic conditions. Industry standards shall be referred to for guidance on dissolved gas concentrations. The laboratory study may not perfectly replicate the variations and concentrations of dissolved gases present in different real-world oil flows. The results shall be interpreted considering the limitations in replicating all possible dissolved gas conditions.
- Flow temperature:
  The flow temperature shall be controlled and monitored to simulate the range of temperatures encountered in oil pipelines. The temperature fluctuations, seasonal variations, and operational conditions that affect the pipeline's corrosion behavior shall be considered. Recognized guidelines shall be followed for determining flow temperature ranges. The laboratory study may not perfectly replicate the temperature fluctuations, seasonal variations, and operational conditions observed in real pipeline systems. The results shall be interpreted considering the limitations in replicating all possible temperature conditions.
- Flow pressure:
  A consistent flow pressure shall be maintained within the corrosion cell to replicate the pressure conditions experienced in actual oil pipelines. It shall be ensured that the pressure is within the safe operating limits for the equipment used. Industry standards shall be followed for guidance on flow pressure limits. The laboratory study may not perfectly replicate the full range of pressure variations and transient events observed in real-world pipeline operations. The results shall be interpreted considering the limitations in replicating all possible pressure conditions.
- Percentage of water:
  The appropriate percentage of water in the synthetic oil solution shall be determined based on the expected water content in the oil flowing through the pipelines. Factors such as emulsions, water ingress, and water separation in the pipeline system shall be considered. Industry standards shall be referred to for guidance on water content in oil. The laboratory study may not fully capture the variations in water content and distribution observed in different real-world pipeline systems. The results shall be interpreted considering the limitations in replicating all possible water conditions.

- Properties of oil:
  The relevant properties of the oil, such as viscosity, density, acidity, and corrosiveness, shall be considered. These properties shall be replicated as closely as possible in the synthetic oil solution to accurately evaluate the corrosion inhibition effectiveness of the imidazoline inhibitors. Industry standards such as ASTM D445 for viscosity measurement or ASTM D5002 for acidity measurement shall be followed. The synthetic oil solution used in the laboratory study may not perfectly replicate all the properties and variations observed in real oil flows. The results shall be interpreted considering the limitations in replicating all possible oil properties accurately.

1.8 Field Testing Procedure for Imidazoline Inhibitors in Oil Pipelines

Field testing of imidazoline inhibitors in oil pipelines involves a systematic and comprehensive procedure to evaluate their effectiveness in preventing corrosion and minimizing the risk of pipeline failure. Outlined below are the steps involved in conducting field tests.

- Planning, preparation, and site selection:
  Prior to conducting the field test, a thorough planning phase is essential. This includes identifying the target pipeline for testing, obtaining necessary permits, and ensuring compliance with safety regulations. The specific sections of the pipeline to be tested should be selected, taking into consideration factors such as pipeline age, location, and corrosion vulnerability. A detailed test plan is developed, specifying the objectives, methods, and parameters to be measured during the test. The planning phase shall be carried out, incorporating industry codes and standards such as API 570 [American Petroleum Institute, 2016A] (Piping Inspection Code: In-Service Inspection, Rating, Repair, and Alteration) and API 653 [American Petroleum Institute, 2018A] (Tank Inspection, Repair, Alteration, and Reconstruction). These codes provide guidelines for the selection of the target pipeline, the acquisition of permits, and compliance with safety regulations. Additionally, API RP 571 [American Petroleum Institute, 2011] (Damage Mechanisms Affecting Fixed Equipment in the Refining Industry) shall be utilized to assess the corrosion vulnerability of the pipeline.
- Selection and application of imidazoline inhibitors:
  Imidazoline inhibitors are selected based on their compatibility with the pipeline material and the expected corrosive environment.

The inhibitors are typically available in liquid form and are applied by injecting them into the pipeline through designated injection points. The dosage and injection frequency are determined based on the inhibitor manufacturer's recommendations and the pipeline characteristics. The selection and application of imidazoline inhibitors shall be performed in accordance with industry standards such as NACE SP0198 [NACE International, 2016A] (Control of Corrosion Under Thermal Insulation and Fireproofing Materials - A Systems Approach) and NACE RP0204 [NACE International, 2017A] (Stress Corrosion Cracking [SCC] Direct Assessment Methodology). These standards provide guidance on inhibitor compatibility, determination of dosage, and frequency of injection.

- Baseline data collection:
  Before initiating the inhibitor application, a comprehensive baseline assessment is conducted to establish the initial condition of the pipeline. This includes measuring corrosion rates, conducting visual inspections, and collecting relevant data such as temperature, pressure, and flow rates. This information serves as a reference for comparing the effectiveness of the inhibitors during subsequent stages of the test. Baseline data collection shall follow the guidelines outlined in API 571 and NACE SP0169 [NACE International, 2015] (Control of External Corrosion on Underground or Submerged Metallic Piping Systems). These standards define parameters for the measurement of corrosion rates, conduct of visual inspections, and collection of data on temperature, pressure, and flow rates.
- Inhibitor application and monitoring:
  Once the baseline data is collected, the imidazoline inhibitors are introduced into the pipeline according to the predetermined dosage and injection schedule. Monitoring and data collection are conducted at regular intervals to assess the inhibitors' performance. This involves measuring corrosion rates, conducting non-destructive testing (e.g., ultrasonic inspection), and periodically inspecting the pipeline's interior for signs of corrosion or inhibitor effectiveness. The process of inhibitor application and monitoring shall be carried out in accordance with API RP 970 [American Petroleum Institute, 2018B] (Corrosion Control Documents) and NACE RP0490 [NACE International, 2012A] (Holiday Detection of Fusion-Bonded Epoxy External Pipeline Coatings of 250 to 760 μm (10 to 30 mils)). These standards outline best practices for the injection of inhibitors into the pipeline, monitoring of corrosion rates, conduct of non-destructive testing, and inspection of the pipeline for corrosion or inhibitor effectiveness.

- Environmental and operating conditions monitoring:
  Throughout the field test, environmental and operating conditions are continuously monitored to evaluate their influence on inhibitor performance. Parameters such as temperature, humidity, pH levels, and the presence of corrosive substances in the oil are recorded. This information helps assess the inhibitors' ability to withstand varying environmental and operational factors. Monitoring of environmental and operating conditions shall follow the guidelines presented in API RP 574 [American Petroleum Institute, 2016B] (Inspection Practices for Piping System Components) and NACE SP0285 [NACE International, 2017B] (Internal Corrosion Direct Assessment Methodology for Pipelines Carrying Normally Dry Natural Gas). These standards define parameters such as temperature, humidity, pH levels, and the presence of corrosive substances in the oil that shall be continuously monitored during the test.
- Analysis and evaluation:
  The collected data is analyzed and evaluated to determine the effectiveness of the imidazoline inhibitors. Statistical techniques, corrosion modeling, and data visualization tools may be employed to interpret the results accurately. The corrosion rates and inhibitor performance are compared against the baseline data to assess the inhibitors' corrosion inhibition efficiency. Data analysis and evaluation shall be performed following API RP 579 [American Petroleum Institute, 2016C] (Fitness-for-Service) and NACE TM0284 [NACE International, 2016B] (Evaluation of Pipeline and Pressure Vessel Steels for Resistance to Hydrogen-Induced Cracking). These standards provide methods for the interpretation of data, conduct of statistical analysis, and modeling of corrosion rates to assess the effectiveness of the imidazoline inhibitors.
- Reporting and recommendations:
  A comprehensive report is prepared, summarizing the field test procedures, results, and conclusions. The report includes detailed findings, statistical analysis, inhibitor performance evaluations, and recommendations for future inhibitor application or modifications to the pipeline's corrosion prevention strategy. The report may also identify areas requiring further investigation or additional tests.
- Follow-up and maintenance:
  Based on the field test results, appropriate follow-up actions are initiated. This may involve adjustments to the inhibitor application strategy, implementing corrosion control measures, or conducting additional tests to validate findings. Regular maintenance and monitoring of the pipeline continue to ensure ongoing corrosion protection and optimize inhibitor performance. Follow-up actions and maintenance activities shall be

conducted following API RP 574 [American Petroleum Institute, 2016B], API RP 579 [American Petroleum Institute, 2016C], and NACE RP0102 [NACE International, 2012B] (In-Line Inspection of Pipelines). These standards guide adjustments to the inhibitor application strategy, implementation of corrosion control measures, and regular maintenance to ensure ongoing corrosion protection and optimization of inhibitor performance.

The specific procedures and protocols for field testing imidazoline inhibitors may vary depending on the pipeline's characteristics, regulatory requirements, and industry standards. Engaging with corrosion experts, following manufacturer recommendations, and adhering to applicable guidelines is crucial throughout the field testing process to ensure accurate evaluation of inhibitor effectiveness and long-term pipeline integrity.

2

Chemistry of Imidazoline Inhibitors

Imidazoline inhibitors have gained widespread recognition for their exceptional ability to counteract CO_2-induced corrosion in oil pipelines. The secret behind their effectiveness lies in a complex interplay of chemical factors. By understanding the chemistry of these inhibitors, we can uncover the intricate mechanisms that underpin their corrosion inhibition properties and explore the diverse range of characteristics they possess. At the core of imidazoline inhibitors lies their distinct molecular structure. These compounds are characterized by the presence of an imidazoline ring, which is formed through a series of chemical reactions involving the combination of aldehydes or ketones with fatty acids or amines. This synthetic process gives rise to various imidazoline derivatives with different lengths of alkyl chains and substitutions of functional groups. This versatility enables customization, allowing for the fine-tuning of their performance.

2.1 Chemical Composition and Characteristics (Preparation and Types)

Imidazoline inhibitors, known for their effective corrosion inhibition properties, possess a diverse range of chemical compositions and characteristics that contribute to their effectiveness in combating CO_2-induced corrosion. Understanding the chemical composition and unique attributes of these inhibitors is crucial for optimizing their performance and tailoring them to specific operating conditions. They are synthesized through a series of chemical reactions involving the reaction of fatty acids or amines with aldehydes or ketones. This synthetic process leads to the formation of imidazoline derivatives, which feature a distinctive imidazoline ring structure.

Imidazoline Inhibitors for Corrosion Protection of Oil Pipeline Steels: Experimental Laboratory Evaluation and Case Studies, First Edition. Faysal Fayez Eliyan.
© 2024 John Wiley & Sons, Inc. Published 2024 by John Wiley & Sons, Inc.

The resulting inhibitors exhibit a delicate balance between hydrophobic and hydrophilic properties, making them amphiphilic in nature. This amphiphilic characteristic enables imidazoline inhibitors to interact with both the hydrophobic metal surface and the surrounding aqueous phase. The alkyl chain length of imidazoline inhibitors plays a significant role in their solubility and effectiveness. Shorter alkyl chains improve solubility in aqueous solutions, facilitating their dispersion and distribution within the corrosive environment. By selecting inhibitors with optimal chain lengths, corrosion engineers can ensure efficient inhibitor performance, even in highly aqueous environments. Furthermore, the substitution pattern and functional groups on the imidazoline ring allow for customization of inhibitor properties. For example, introducing functional groups with electron-donating or electron-withdrawing properties can influence the inhibitor's adsorption behavior and its interaction with the metal surface.

In addition to their chemical composition, imidazoline inhibitors exhibit other noteworthy characteristics. They possess surface active properties, enabling them to adsorb onto the metal surface and form a protective film. This film acts as a physical barrier, preventing direct contact between the metal and corrosive species, such as CO_2 and corrosive ions present in the environment. Imidazoline inhibitors also demonstrate excellent film-forming abilities, allowing them to create a continuous, homogeneous layer on the metal surface. This layer effectively hinders the diffusion of CO_2 and corrosive ions, impeding their access to the underlying metal substrate and reducing the corrosion rate. The formation of a stable, less corrosive corrosion product layer is facilitated by the strong adsorption of imidazoline inhibitors on the metal surface, promoting the passivation process.

The molecular structure of an imidazoline-based inhibitor can be divided into three different substructures: a five-atom ring with nitrogen elements, a long hydrocarbon chain (R_1), and a pendant side chain with an active functional group (R_2), shown in Figure 2.1, which are functional groups that could be variables. R_1 proportionately with its hydrocarbon chain length plays at variant degrees a significant role in the inhibition performance [Ramachandran, 1996, Jovancicevic, 1998, Cruz, 2004]. R_2 was reported to control the inhibition performance and adsorption at different degrees [Edwards, 1994, Wang, 1999, Cruz, 2004].

In general, the range of use of imidazolines, as organic nitrogen-based surfactants to mitigate corrosion in oil and gas industry, was reported to inhibit the anodic reactions more than the cathodic reactions [Moiseeva, 1997]. They have good adsorption characteristics, and form stable protective films on inner pipeline steel surfaces, used in forms

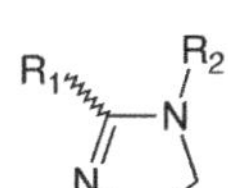

Figure 2.1 Schematic of a typical imidazoline molecule.

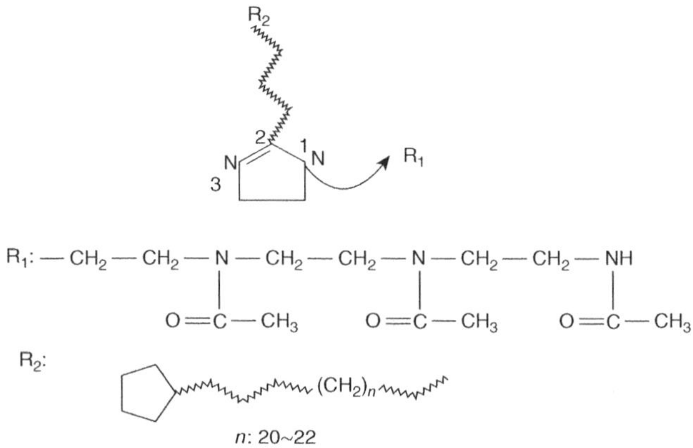

Figure 2.2 Molecular structure of imidazoline amide derivative. *Source:* Zhang [2009].

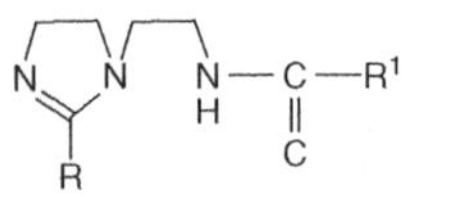

Figure 2.3 Molecular structure of imidazoline amido amine derivative. *Source:* Usman [2018].

of amides, amidoamines, and in salt forms. The molecular structure of imidazoline amide, in Figure 2.2, and that of imidazoline amidoamine, in Figure 2.3, consists of three parts [Zhang, 2009]: a five-atom ring with nitrogen element (part A), a chain with an active functional group (R_1, part B) and a long hydrocarbon chain (R_2, part C). The functional groups R_1 and R_2 are variable.

Common imidazoline inhibitors include polymeric inhibitors such as polyimidoamines and polypeptides [Tiu, 2015]. They have excellent corrosion inhibition performance in acidic and high-temperature environments. The polyimidoamines are produced with a polyamine reacting with an anhydride copolymer in a condensation environment. The polyamines enhance the inhibitor film durability; they commonly include 1,4-butanediamine (H_2N-$(CH_2)4$-NH_2), ethylenediamine (H_2N-CH_2-CH_2-NH_2), trimethylenediamine ($C_3H_{10}N_2$), and tetraethylenepentamine (H_2N-(CH_2-CH_2-NH)4H). Polypeptides are produced with similar procedures and, attributed to the calcium and high pH, a polypeptide such as polyaspartate has a high inhibition performance. An example of imidazoline inhibitor which has tall oil fatty acids with oleic acid as a major component is tall oil diethylenetriamine (TOFA/DETA) imidazoline, with chemical structure shown in Figure 2.4. It consists of three different substructures: a nitrogen-containing five-member ring, a pendant side chain with a

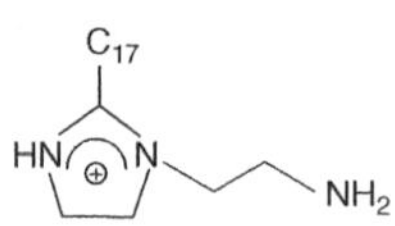

Figure 2.4 Molecular structure of tall oil diethylenetriamine imidazoline (TOFA/DETA imidazoline). *Source:* Jevremović [2016].

hydrophilic active functional group (ethylamino pendent group) that is attached to N1 and a long C-17 hydrophobic chain attached to a C_2 atom. The imidazoline-typeinhibitor,1-(2-aminoethyl)-2-heptadecyl imidazoline (IM, CAS: 3010-23-9), used by Li *et al.* was of the molecular structure shown in Figure 2.5 [Li, 2017].

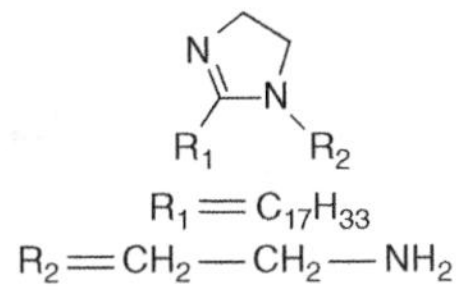

Figure 2.5 Molecular structure of 1-(2-aminoethyl)-2-heptadecyl imidazoline inhibitor.

Imidazoline derivatives of good efficiency can be synthesized from, in addition to oleic acid and thiourea, reactions that involve diethylenetriamine, as illustrated in Eqs. (1.1)–(1.3) [Chai, 2005, Zhang, 2007]:

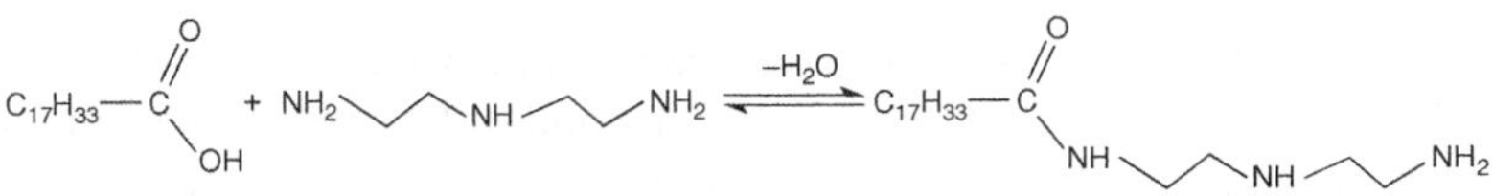

(2.1) [Chai, 2005]

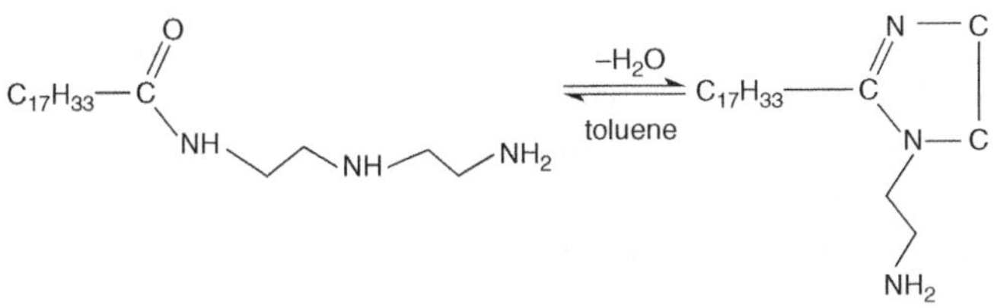

(2.2) [Chai, 2005]

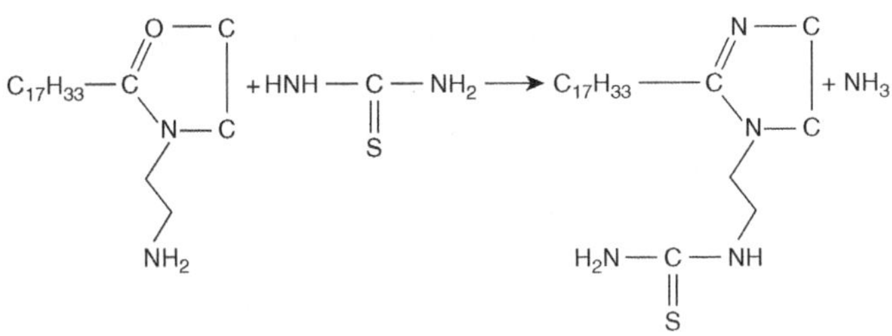

(2.3) [Chai, 2005]

Shamsa *et al.* synthesized their imidazoline inhibitor, of molecules shown in Figure 2.6, from the reaction of tall oil fatty acid (TOFA) and diethylenetriamine (DETA) [Shamsa, 2020]. Singh *et al.* synthesized three imidazoline inhibitors 1,3-bis(4-methoxybenzyl)-2-(4-methoxyphenyl)

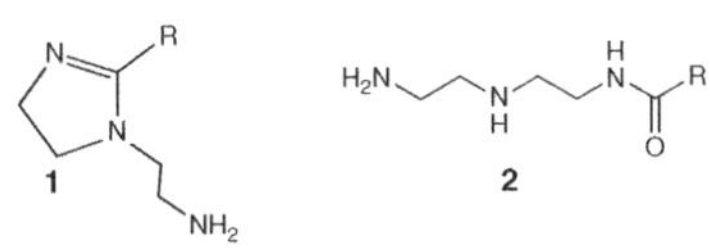

Figure 2.6 The main components of imidazoline based DETA/TOFA corrosion inhibitor: (1) imidazoline, and (2) amide. *Source:* Shamsa [2020].

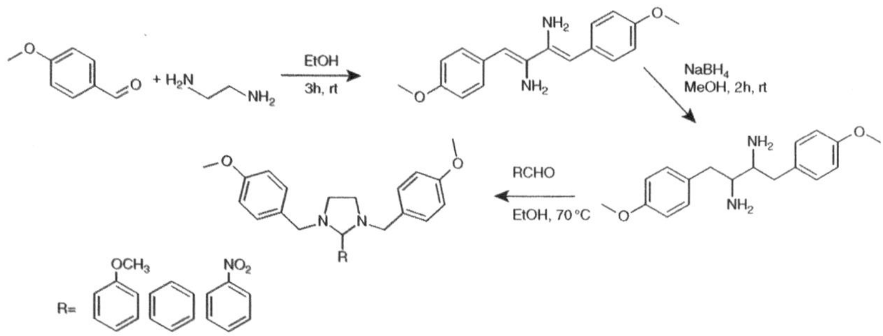

Figure 2.7 Synthetic scheme of AT inhibitors. *Source:* Singh [2019].

imidazoline (AT-1), 1,3-bis (4-methoxybenzyl)-2-phenylimidazolidine (AT-2), 1,3-bis(4-methoxybenzyl)-2-(4-nitrophenyl) imidazoline (AT-3), according to the procedure shown in Figure 2.7, and characterized ^{1}H NMR and ^{13}C NMR spectra, with the molecular structures and analytical data in Table 2.1 [Singh, 2019].

Ortega-Toledo *et al.* used commercial grade carboxyethylimidazoline, from Lakeland Labs (UK), with the molecular structure in Figure 2.8 [Diaz, 2010, Ortega-Toledo, 2010]. It was composed of a five-member ring containing nitrogen elements, a C-14 saturated hydrophobic head group, and a pendant hydrophilic carboxy group attached to one of the nitrogen atoms. For tests, the authors dissolved the inhibitor in pure 2-propanol. Chen *et al.* synthesized imidazoline by using oleic acid and diethylenetriamine, with the pathway shown below in Figure 2.9, and the composition was verified by Fourier-transform infrared spectroscopy (FTIR) [Chen, 2015]. Solomon *et al.* examined the corrosion inhibition of an imidazoline, 2-hept adecyl-1-[2-(octadecanoylamino)ethyl]-2-imidazoline (QSI) with -$C_{17}H_{35}$ as the tail chain length was synthesized, characterized, and studied as a corrosion inhibitor, shown in Figure 2.10 [Solomon, 2020]. The chemicals used for the synthesis of QSI include diethylenetriamine (99%), stearic acid (>97%), calcium oxide (>99.99%), and ethyl acetate (99.8%).

Tantawy *et al.* managed to prepare three cationic surfactants with various isolated yields ranging from 75 to 86%, shown in Figure 2.11 [Tantawy, 2020]. The imidazolines inhibitors included *N*-(3-((2*E*,4*E*)-5-(benzo[d][1,3]dioxol-5-yl)penta-2,4-dienamido) propyl)-*N*,*N*-dimethyloctan-1-aminium iodide (PGS-8), *N*-(3-((2*E*,4*E*)-5-(benzo[d][1,3]dioxol-5-yl)penta-2,4dienamido) propyl)-*N*,*N*-dimethyldecan-1-aminium iodide (PGS-10) and *N*-(3-((2*E*,4*E*)-5-(benzo[d][1,3]dioxol-5-yl)penta-2,4-dienamido)propyl)-*N*,*N*-dimethyldodecan-1-aminium iodide (PGS-12). The procedure involved alcoholic KOH, piperic acid, H_2SO_4, and derivatives in ethyl acetate as a solvent.

Table 2.1 Molecular structure and analytical data of AT inhibitors.

Inhibitor	Structure	Analytical data
1,3-Bis(4-methoxybenzyl)-2-(4-methoxyphenyl)imidazolidine (AT-1)	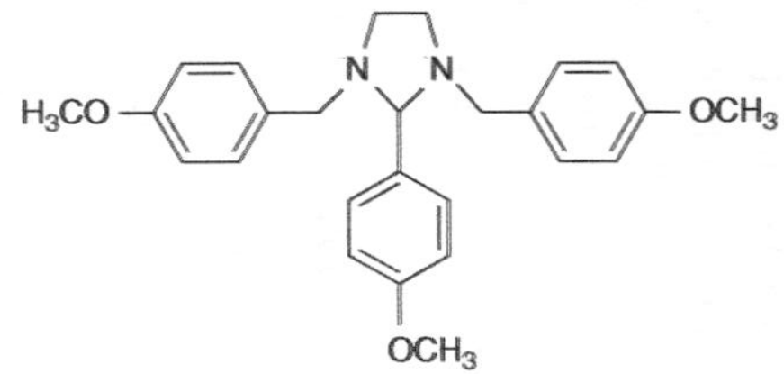	M.P. 90–92 °C; ^{1}**H NMR** (300 MHz, DMSO-d_6) δ (ppm): 2.46–2.54 (m, 4H), 3.67 (s, 2H), 3.82 (s, OCH_3), 5.45 (s, 1H), 6.60–7.35 (m, 12H); ^{13}**C NMR**, δ (ppm): 48.23, 55.09, 66.15, 69.25, 91.21, 113.60, 125.94, 129.15, 131.71, 140.55, 154.68, 157.97, 158.99, 161.51
1,3-Bis (4-methoxybenzyl)-2-phenylimidazolidine (AT-2)	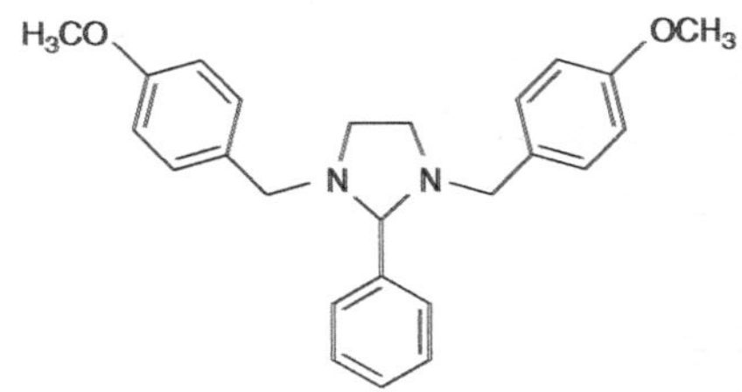	M.P. 74–76 °C; ^{1}**H NMR** (300 MHz, DMSO-d_6 δ 2.34–2.46 (m, 4H), 3.67 (s, 2H), 3.83 (s, OCH_3), 5.47 (s, 1H), 6.71–7.34 (m, 13H) ^{13}**C NMR** δ (ppm): δ 48.79, 55.13, 68.72, 90.81, 113.63, 125.99, 131.94, 140.70, 154.76, 158.01, 162.00
1,3-Bis(4-methoxybenzyl)-2-(4-nitrophenyl)imidazolidine (AT-3)	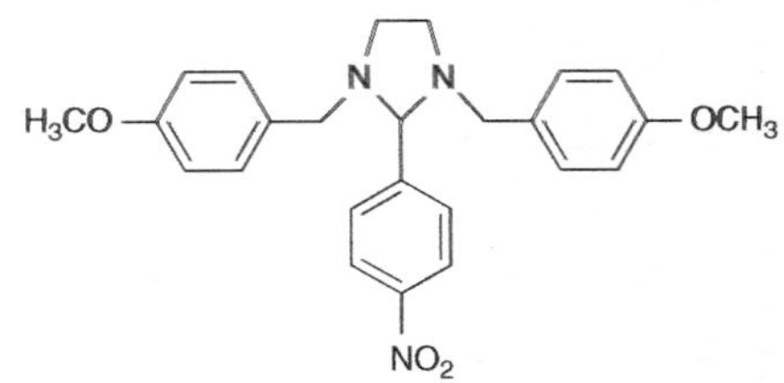	M.P. 148–150 °C; ^{1}**H NMR** (300 MHz, DMSO-d_6) δ 2.46–2.56 (m, 4H), 3.68 (s, 2H), 3.84 (s, OCH_3), 5.54 (s, 1H), 6.76–7.52 (m, 10H), 8.01–8.12 (m, 2H) ^{13}**C NMR** δ (ppm): 48.76, 56.78, 68.66, 90.77, 116.81, 125.17, 132.71, 140.79, 143.07, 147.71, 154.85, 161.67

Source: Singh [2019]/with permission of Elsevier.

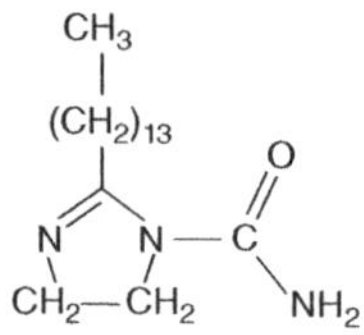

Figure 2.8 Molecular structure of carboxyethyl-imidazoline. *Source:* Diaz [2010].

Zhang *et al.* synthesized an imidazoline inhibitor from stearic acid, diethylenetriamine, and thiourea in a 250-ml three-neck round-bottom flask, with a magnetic rotor, and a condenser to collect water and xylene [Zhang, 2015]. Under the protection of nitrogen, 22.74 g stearic acid, 10.3 ml diethylenetriamine, and 24 ml xylene were heated as a mixture at between 155 and 165 °C, stirred for three hours, then heated at between 195 and 220 °C, and stirred for six hours. The imidazoline intermediate was obtained after that process after removing the abundant xylene and water. 7.4 g of thiourea was added to the intermediate, with octyl alcohol as solvent, at between 140 and 160 °C with stirring for five hours. The redundant octyl alcohol was removed by reduced pressure distillation, and the final products was the inhibitor of the molecular structure in Figure 2.12.

The characteristics of the inhibitor were evaluated with FTIR spectroscopy in the range of 4000 to 5000 cm^{-1}, were FTIR (KBr) ν: 3299 (N—H), 2923 and 2853(—CH_2—), 1610 (C=N), 1465 (—CH_3), 1301 (C—N), 1053 (C=S), 721(—CH_3). The melting point of the synthesized inhibitor, of good solubility in water, was between 73 and 76 °C. The inhibitor, in different

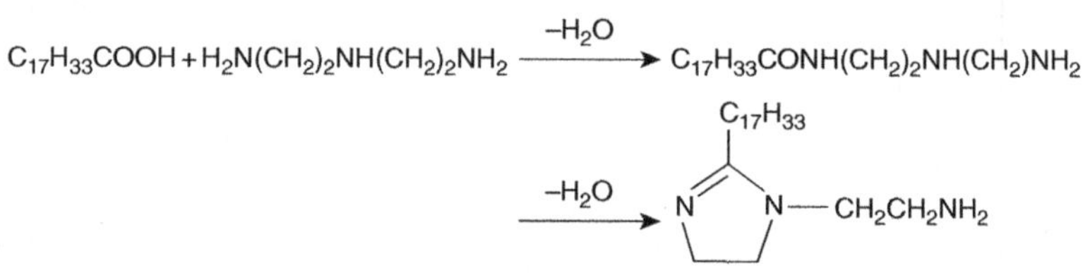

Figure 2.9 Reaction pathway of oleic acid with diethylenetriamine. *Source:* Chen [2015].

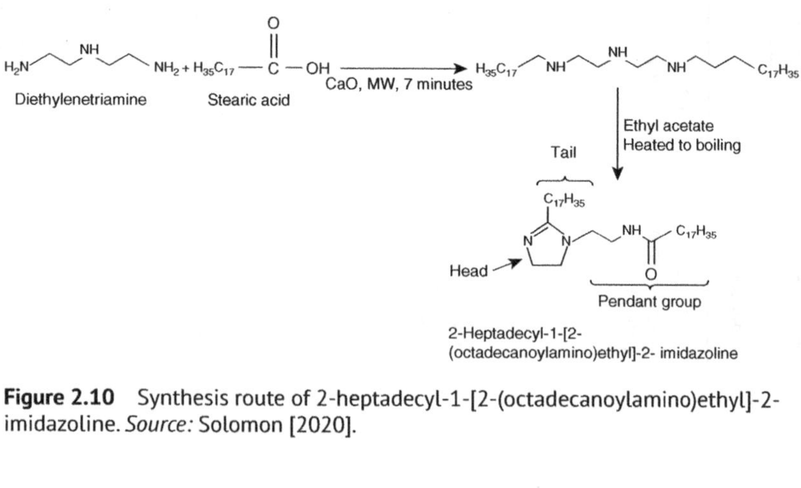

Figure 2.10 Synthesis route of 2-heptadecyl-1-[2-(octadecanoylamino)ethyl]-2-imidazoline. *Source:* Solomon [2020].

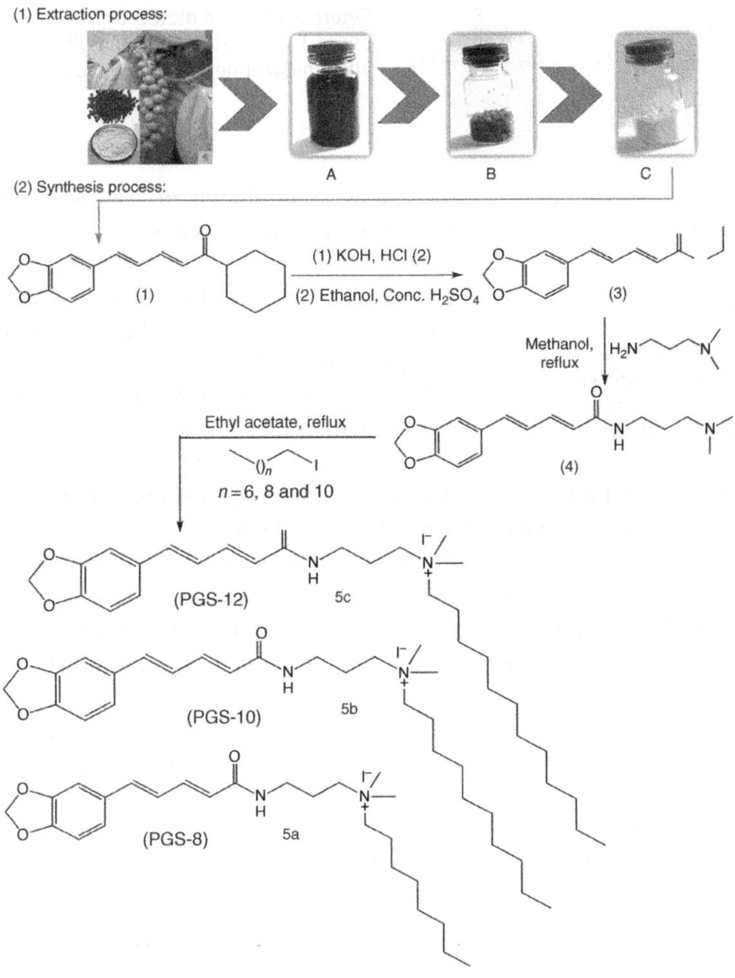

Figure 2.11 Scheme 1. Synthetic route for the preparation of the novel cationic surfactants (PGS-8, PGS-10, PGS-12). *Source:* Tantawy et al., [2020]/with permission from Elsevier.

concentrations, that Wang *et al.* tested with the corrosion of Q235 steel in CO_2-saturated solutions was 1-(2-thioureidoe thyl)-2-alkylimidazoline (TAI), of the molecular structure shown in Figure 2.13 [Wang, 2009]. R represents alkyl C17H35, which was confirmed by FTIR analysis, by Zhao *et al.* [Zhao, 2018].

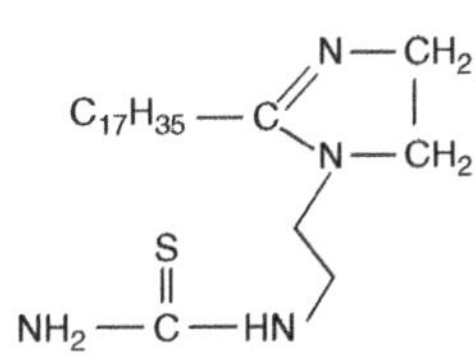

Figure 2.12 Molecular structure of the synthesized inhibitor. *Source:* Zhang [2015].

The molecular structure of 2-undecyl-1-ethylamino imidazoline (2UEI) tested

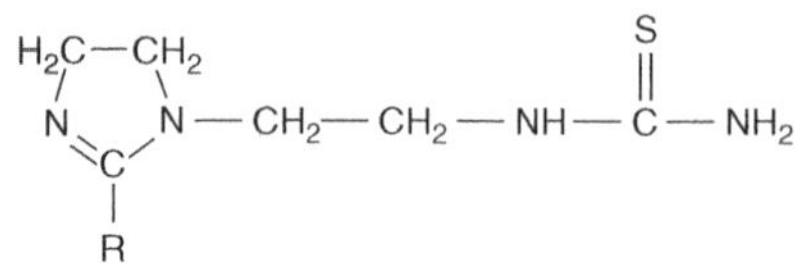

Figure 2.13 The molecular structure of the TAI (1-(2-thioureidoethyl)-2-alkylimidazoline). *Source:* Wang [2009].

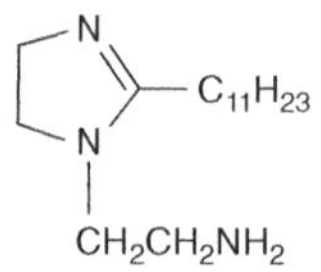

Figure 2.14 Molecular structure of 2-undecyl-1-ethylamino imidazoline (2UEI). *Source:* Chen [2019].

for corrosion inhibition of N80 steel by Okafor *et al.* was composed of a five-member ring containing nitrogen elements (the head group), a C-11 saturated hydrophobic group, and a hydrophilic ethylamino group attached to N1, shown in Figure 2.14 [Chen, 2019, Okafor, 2009]. The structure was similar to the two inhibitors used by Liu *et al.* 2-undecyl-1-aminoethyl imidazoline (AEI-11) and 2-undecyl-1-aminoethyl-1-hydroxyethyl quaternary imidazoline (AQI-11), shown in Figure 2.15 [Liu, 2009A].

Belarbi *et al.* used the inhibitor 1-(2-aminoethyl)-2-oleyl-2-imidazolinium of the long-chain molecular structure, in Figure 2.16, for inhibiting corrosion in 1% NaCl solutions of controlled pH 4 [Belarbi, 2017], and Olivares *et al.* used oleic imidazoline, in Figure 2.17, for inhibiting corrosion of a

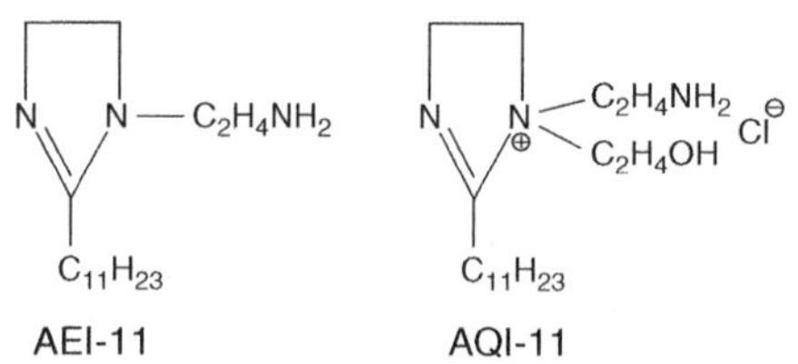

Figure 2.15 Molecular structures of AEI-11 and AQI-11. *Source:* Liu [2009A].

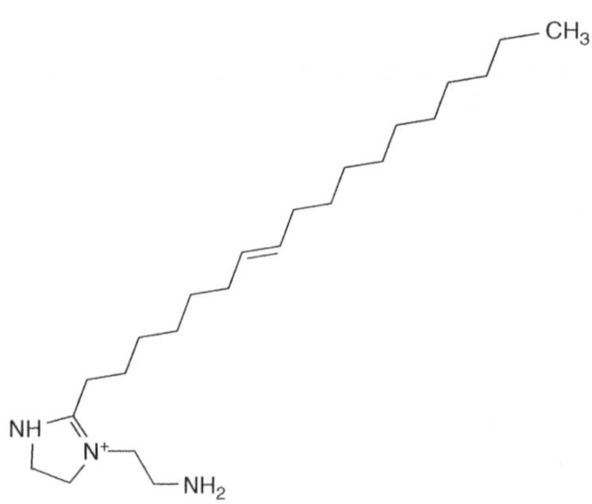

Figure 2.16 Molecular structure of 1-(2-aminoethyl)-2-oleyl-2-imidazolinium chloride. *Source:* Belarbi [2017].

mild steel in 3% NaCl solutions of pH values of 3.8 and 5.5 [Olivares, 2007].

Rivera-Grau *et al.* used organic hydroxyethyl imidazoline (HEI) inhibitor, made by Lakeland Labs (UK), with a molecular structure in Figure 2.18, introduced in 10 ppm, to inhibit corrosion in CO_2-saturated brines containing diesel [Rivera-Grau, 2016].

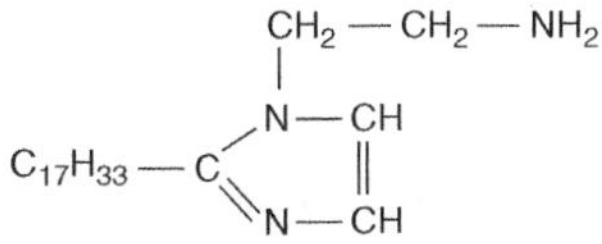

Figure 2.17 Molecular structure of oleic imidazoline. *Source:* Zavala Olivares [2007].

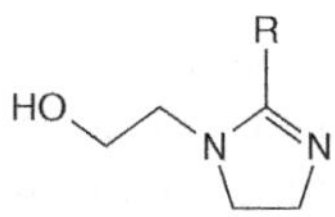

Figure 2.18 Chemical structure of hydroxyethyl-imidazoline (HEI), where R is an ethyl group. *Source:* Rivera-Grau [2013].

Ortega-Sotelo *et al.* used commercial carboxyamido imidazoline in the corrosion evaluation of X70 steel [Ortega-Sotelo, 2011]. In the molecular structure shown in Figure 2.19, the inhibitor, dissolved in 100 ml pure 2-propanol in experiments, consists of a five-member ring containing nitrogen elements, a C-14 saturated hydrophobic head group, and a pendant, hydrophilic carboxyamido group attached to one of the nitrogen atoms.

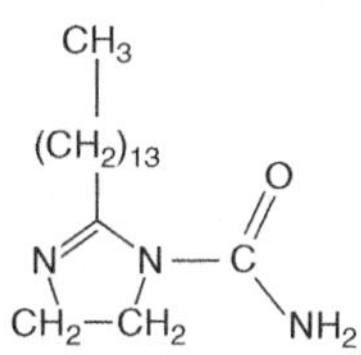

Figure 2.19 Molecular structure of carboxyamido imidazoline. *Source:* Ortega-Sotelo [2011].

Other inhibitors are composed of polymerized amides in a mixture of ammonium-based salts, long-chain fatty acids, resin acids, and dispersing agents. A typical process of preparing amide inhibitors involves converting methyl methacrylate with tallow triamine or tetramine at around 80 °C to a corresponding amide. After that, the amide is polymerized at higher temperatures to produce the polymer agent that interferes with the corrosion process and form the protective film. Some amides, converted for example from ethoxylated alkylphenol amines, contain fatty acids that effectively interfere and control both CO_2 and H_2S corrosion. Such fatty acids include oleic acid ($C_{18}H_{34}O_2$), myristic acid ($H_3C\text{-}(CH_2)_{12}\text{-}COOH$), stearic acid ($H_3C\text{-}(CH_2)_{16}\text{-}COOH$), palmitic acid ($H_3C\text{-}(CH_2)_{14}\text{-}COOH$) and lauric acid ($H_3C\text{-}(CH_2)_{10}\text{-}COOH$). They have low melting temperatures and they typically contain resin acids, oleic acid, and a mixture of other acids with a saturated acid at low proportions. Some amide inhibitors include ammonium salts that are half-amides of, for example, alkenyl succinic, and/or other sorts that inhibit specific corrosion types in oil pipelines. The amide inhibitors disperse in the multiphase flows with surfactants that are polymeric and/or anionic, like alkyl aryl sulfonic acids.

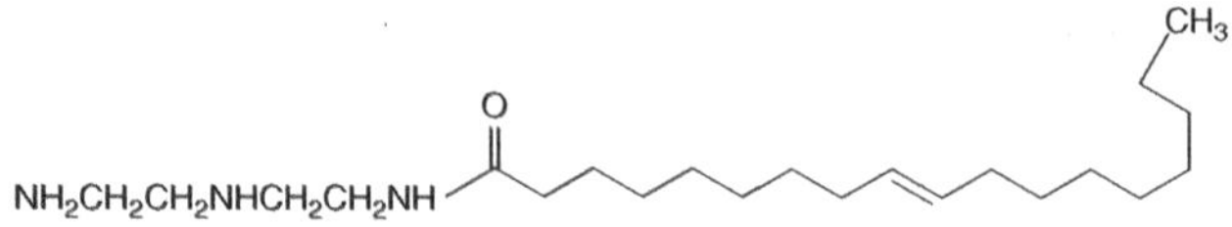

Figure 2.20 Molecular structure for *N*-[2-[(2-aminoethyl) amino] ethyl]-9-octadecenamide. *Source:* Jiang [2013].

Desimone *et al.* synthesized *N*-[2-[(2-aminoethyl) amino] ethyl]-9-octadecenamide (AAOA), with molecular structure shown in Figure 2.20, from oleic acid and diethylenetriamine, according to a detailed procedure by Olivares-Xometl *et al.*, for which the inhibitor was used also by Jiang *et al.* [Olivares-Xometl, 2006, Desimone, 2011, Jiang, 2013].

Yan *et al.* synthesized modified amide inhibitor according to a novel procedure detailed in [Zou, 2011] by Zou et al. The inhibitor was water-soluble polyacrylamide (PAM) modified by b-cyclodextrin (ACDPAM) by introducing b-CD group in conjunction with dimethyl diallyl ammonium chloride. The molecular structure is shown in Figure 2.21.

Cyclodextrins (CDs; see Figure 2.22) are cyclic oligosaccharides consisting of six to eight glucose units linked by 1,4-a-glucosidic bonds, whereas the b-CD containing seven glucose units has more practical significance. The primary hydroxyl groups and secondary hydroxyl groups in b-CD form a complex network of intramolecular hydrogen bonds to give a polar hydrophilic outer shell and a relatively hydrophobic cavity [Li, 1992]; thus, b-CD can form host/guest inclusion complexes with organic compounds.

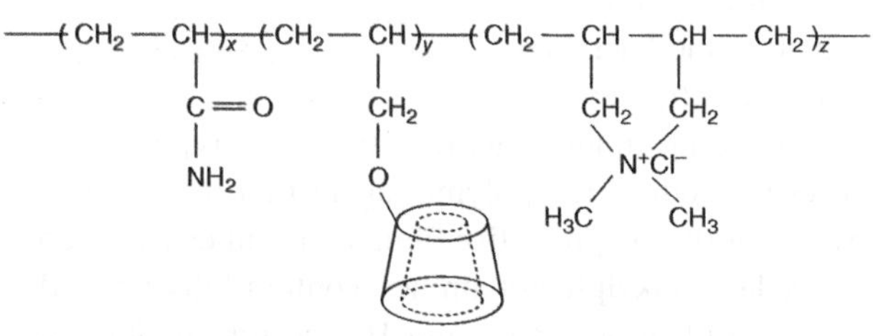

Figure 2.21 Molecular structure of ACDPAM. *Source:* Yan [2014].

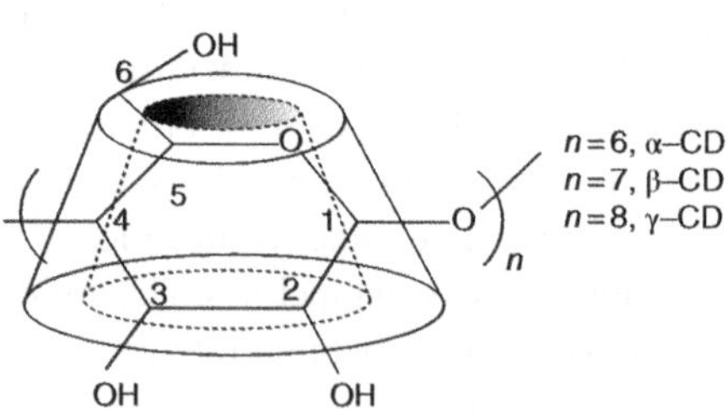

Figure 2.22 Structure of CDs. *Source:* Zou [2011].

To enhance performance of b-CD into PAM, the novel water-soluble polymer acrylamide/dimethyl diallyl ammonium chloride/allyl-b-CD (AM/DMDAAC/allyl-b-CD) was designed, of synthesis process illustrated [Zou, 2011]:

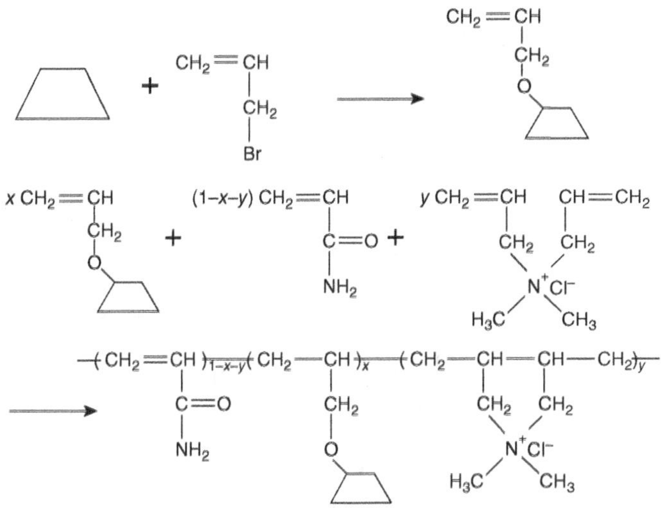

(2.4) [Zou, 2011]

Okafor *et al.* studied the inhibition of corrosion of two carbon steels by rosin amide imidazoline (RAIM), of the structure shown in Figure 2.23 [Okafor, 2011]. From the electrochemical analysis they reported that it chemically adsorbs onto the surface by charge-sharing or charge-transfer to form a coordinate-type bond. The molecular structure consists of a five-member ring containing nitrogen elements (the head group), an unsaturated hydrophobic group, and a hydrophilic $C_2H_5NH_2$-group attached to the N1 atom. The compound adsorbs on the metal surface as a neutral molecule by an iron–nitrogen (Fe—N) coordination bond, and by a π-electron interaction between the π-electron in the head group and iron, or the compound can adsorb in the cationic form by the positively charged part (ammonium-NH_2) of the particle, oriented toward the negatively charged surface. The adsorbed film protects the surface from mass transfer and charge transfer for corrosion.

Migahed *et al.* synthesized 1-(2-aminoethyl)-1-dodecyl-2-hexadecyl-4,5-dihydro-1H-imidazol-1-ium (PQI) and 1-(2-aminoethyl)-1-dodecyl-2-hept adecyl-4,5-dihydro-1H-imidazol-1-ium (SQI) for the sake of studying corrosion inhibition of carbon steel in deep oil wells formation water [Migahed, 2017]. The authors reported the preparation process to involve

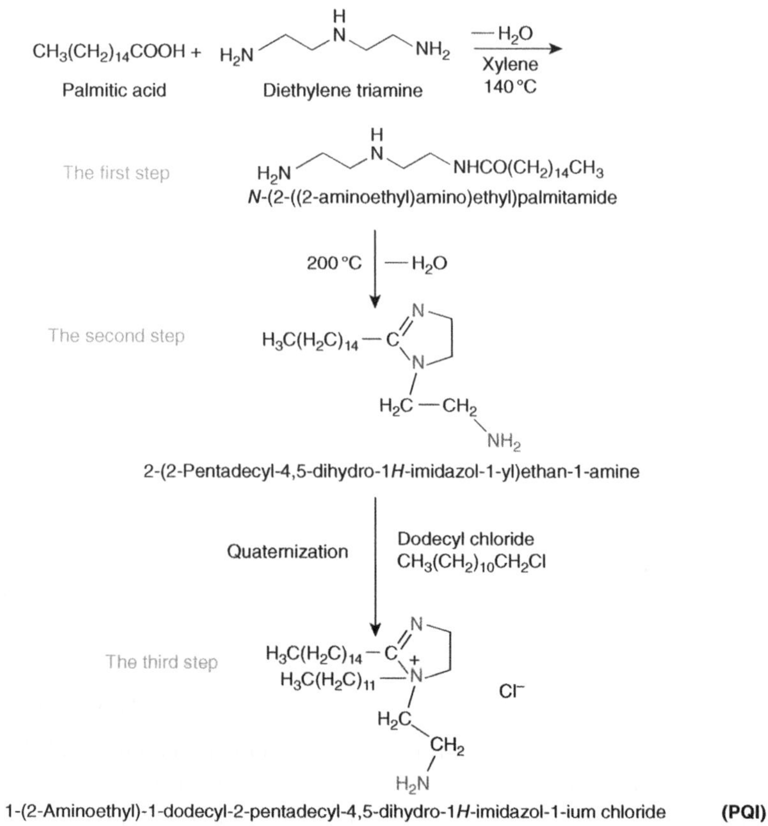

Figure 2.23 Chemical structure of RAIM. *Source:* Okafor [2011].

the addition of 0.1 mol diethylenetriamine drop by drop to monocarboxylic fatty acids, 0.1 mol palmitic, and 0.1 mol stearic, in the presence of 100 ml xylene solvent. This was in a setup of 500-ml three-necked flask, glass condenser, water splitter, and a dropping funnel. In the first step, the mixture was refluxed at 140 °C for three hours, and the calculated amount of 1.8 ml water was obtained. In the second step, xylene was removed by distillation, and the mixture was gradually heated to 200 °C for seven hours, for another calculated amount of 1.8 ml water was obtained, and the mixture was cooled to 70 °C. In the third step, dodecyl chloride was added drop by drop and the mixture was refluxed at 100 °C for 20 hours. The whole synthesis process is illustrated as pattern 1 and pattern 2, for synthesis of PQI compound, and SQI compound, respectively, below in Figure 2.24, and 2.25, respectively.

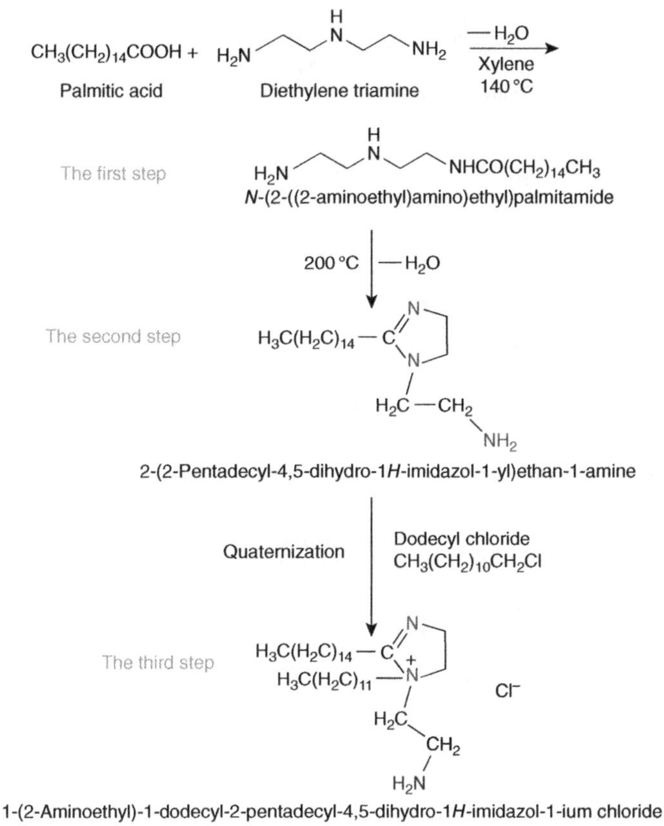

(PQI)

Figure 2.24 Synthesis of PQI compound. *Source:* Migahed [2017].

Lucio-Garcia *et al.* and Ortega-Toledo *et al.* used a commercial type HEI, dissolved in pure 2-propanol and of the chemical structure in Figure 2.26, of a five-member ring containing nitrogen elements, a C-14 saturated hydrophobic head group and a pendant, hydrophilic hydroxyethyl group attached to one of the nitrogen atoms [Ortega-Toledo, 2010].

2.2 Mechanisms of Inhibition

The effectiveness of imidazoline inhibitors in mitigating CO_2-induced corrosion stems from their multifaceted mechanisms of inhibition. One of the primary mechanisms through which imidazoline inhibitors provide corrosion inhibition is by forming a protective film on the metal surface. When

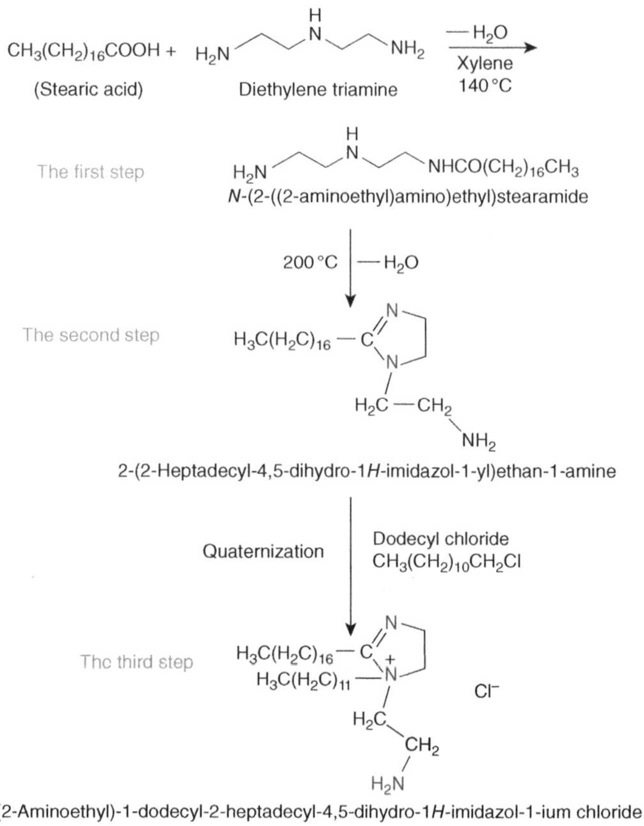

(SQI)

Figure 2.25 Synthesis of SQI compound. *Source:* Migahed [2017].

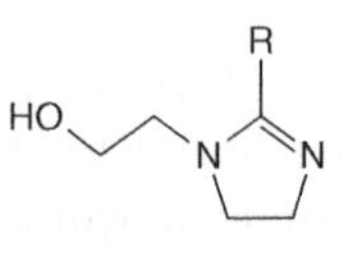

Figure 2.26 General structure of hydroxyethyl imidazoline, where R is an alkyl chain derivative. *Source:* Lucio-Garcia [2010] and Ortega-Toledo [2010].

introduced into the corrosive environment, the inhibitors adsorb onto the metal surface, creating a thin, uniform film that acts as a physical barrier. This film prevents direct contact between the corrosive species and the metal, effectively impeding the corrosion process. The formation of this protective film is facilitated by the strong affinity of imidazoline inhibitors for metal surfaces, which promotes their adsorption and film formation abilities. Imidazoline inhibitors also exhibit passivating properties, contributing to their corrosion inhibition efficacy. Passivation refers to the formation of a stable, less corrosive corrosion product layer on the metal surface. The adsorption of imidazoline inhibitors enhances the formation of this protective oxide layer, reducing the metal's reactivity toward corrosive species. This passivation process is facilitated by the inhibitor's ability to promote the deposition of a stable and adherent corrosion product layer, which acts as an additional barrier against corrosion. Furthermore, imidazoline inhibitors function as neutralizers by interacting with the corrosive species formed by CO_2 dissolution. The amine groups present in the inhibitors react with the acidic species generated in the presence of CO_2, effectively neutralizing the local acidity. This neutralization reaction reduces the corrosive potential of the environment, mitigating the degradation of the metal surface. By acting as alkaline reserve sources, imidazoline inhibitors help maintain a more favorable pH near the metal surface, further inhibiting corrosion.

The inhibitory mechanisms of imidazoline inhibitors are not limited to film formation, passivation, and neutralization. These inhibitors can also modify the electrochemical properties of the metal surface, shifting it to a more passive state. This modification occurs through interactions between the inhibitor molecules and the metal surface, altering the metal's electron transfer processes and reducing its susceptibility to corrosion. Additionally, imidazoline inhibitors may exhibit self-healing properties. In the event of localized damage to the inhibitor film, the inhibitors can redistribute and reform the protective film, effectively repairing the damaged areas. This self-healing behavior contributes to the long-term effectiveness of imidazoline inhibitors in inhibiting corrosion.

By understanding these complex mechanisms of inhibition, corrosion engineers can tailor the formulation and application of imidazoline inhibitors to maximize their corrosion inhibition efficacy. Optimizing the inhibitor concentration, film formation parameters, and interaction with corrosive species allows for the development of highly effective corrosion mitigation strategies in oil pipelines, ensuring the longevity and reliability of the infrastructure.

The modes of the inhibition of organic inhibitors are classified into three categories: geometric blocking by adsorbed inhibitive species (physisorption), active sites blocking by adsorbed inhibitive species (chemisorption), and electro-catalytic effect of the inhibitor or its reaction products. The coverage of corrosion inhibitor of the surface is expressed with (θ), which is calculated as $\theta = \eta/100$; (η) is the inhibition efficiency, which can be calculated from the corrosion rates calculated or measured from the electrochemical or weight loss measurements [Zhang, 2001]. The metal surface immersed in aqueous solutions gets covered with water dipoles, and the adsorption of the organic inhibitors in the aqueous phase ($Org_{(sol)}$) is considered as substitution [Obot, 2009], as:

$$Org_{(sol)} + xH_2O_{(ads)} \leftrightarrow Org_{(ads)} + xH_2O \tag{2.5}$$

Surface coverage θ were fit as values in adsorption isotherms, including Frumkin, Freundlich, Langmuir, and Temkin isotherms [Heydari, 2012].

Langmuir isotherm model is used for the assumption that each site of the metal surface holds an adsorbed species; as one water dipole is replaced by one organic inhibitor molecule, and the adsorbed inhibitor molecules do not interact with each other. Langmuir isotherm is mathematically expressed as $\left(\frac{\theta}{\theta-1}=K_{\text{ads}}C\right)$, where ($K_{\text{ads}}$) is the equilibrium constant of the adsorption reaction. The equation is rearranged to $\left(\frac{C}{\theta}=\frac{1}{K_{ads}}+C\right)$ for plotting (c/θ) against C to produce an adsorption straight line. K_{ads} is related to the standard free energy of adsorption ($\Delta G^{o}_{\text{ads}}$) as:

$$K_{\text{ads}}=\frac{1}{55.5}\exp\left(\frac{\Delta G^{o}_{\text{ads}}}{RT}\right) \tag{2.6}$$

(R) is the universal gas constant (8.3144 J/K mol) and (T) is the temperature in K. The value of 55.5 is the concentration of water in solution expressed in moles per liter. After substituting K_{ads}, the energy of adsorption ($\Delta G^{o}_{\text{ads}}$) can be calculated. The more negative the energy the higher the tendency for adsorption to take place thermodynamically. In general, energy values ranging to up to −20 kJ/mol correspond to physisorption where the inhibitor molecules could well electrostatically interact with the iron surface. When adsorption energy is as high to nearly −40 kJ/mol, the charge transfer between the inhibitor molecules and iron surface becomes more favorable and coordinate bonds for chemisorption form.

Temkin adsorption isotherm expresses an interfacial condition under which the adsorbed inhibitor molecules interact with each other with the mathematical expression:

$$\theta=\frac{1}{f}\ln K_{\text{eq}}+\frac{1}{f}\ln c \tag{2.7}$$

(f) is Temkin heterogeneity factor, (a) is Frumkin interaction parameter. Temkin heterogeneity factor is related to the Frumkin interaction parameter as:

$$f=-2a \tag{2.8}$$

Frumkin interaction parameter (a) is described as: ($a = 0$, no interaction; $a > 0$, attraction; $a < 0$, repulsion), and it is dependent on the hydrophilic head group charge, hydrophobic chain length, flexibility, and branching.

Imidazolines evaluated with electrochemical methods were found to have high inhibition efficiencies in a variety of pipeline conditions. They inhibit both the anodic and cathodic reactions in CO_2-saturated brines. The ring part that has nitrogen, together with the part that has unsaturated double bonds, adsorbs chemically on the surface and forms a stable film [Zhang, 2001]. The nature of the stable film depends on the inhibitor concentration. Zhang *et al.*, from Langmuir's isotherms, postulated that imidazoline inhibitors form a single-layer film, and Jovancicevic *et al.* postulated that with higher concentrations the film becomes of double or multiple layers [Jovancicevic, 1998]. The part that has the hydrocarbon chain maintains a chemical barrier against water. With the use of polarization modulation infrared reflection-absorption spectroscopy (PM-IRRAS), Desimone *et al.* postulated that the adsorption of AAOA occurs by primary or secondary amine and/or amide functional groups [Desimone, 2011]. More specifically, the adsorption occurs as a chemisorption between AAOA molecules and steel surface, by links between the d orbital of iron atoms and sp^2 electron pairs of N and/or O atoms. Zhang *et al.* suggested the mechanism of the adsorption of an imidazoline inhibitor from the electrochemical impedance spectroscopy response of steel in CO_2-saturated brines [Zhang, 2007]. From the phase shift during the immersion they suggested that the inhibitor adsorbs as a bilayer film. They suggested that the inhibitor adsorbs on the surface in the form of a neutral molecule in chemisorption that involves the sharing of electrons between the nitrogen and iron atoms. The N—C—N bond in imidazoline molecules is of p–π conjugation property and the carbon atom has sp^2 hybrid orbital geometry. They mentioned that adsorption can also occur in the cationic form with the positively charged part of the molecule (ammonium – $NH3^+$) oriented toward the negatively charged iron surface.

In rotation experiments with low 8.1×10^{-5} mol/l carboxyethyl-imidazoline, Ortega-Toledo *et al.* suggested that the inhibitor follows the well-accepted mechanism of inhibition at first steps when the inhibitor replaces water molecules on the surface [Ortega-Toledo, 2010]:

$$\mathrm{Inh_{sol}} + x\mathrm{H_2O_{ads}} \rightarrow \mathrm{Inh_{ads}} + x\mathrm{H_2O_{sol}} \tag{2.9}$$

The inhibitor combines with the ferrous ions and forms iron-inhibitor complexes [Jiang, 2005A; Oguzie, 2007]:

$$\mathrm{Fe} \rightarrow \mathrm{Fe^{2+}} + 2\mathrm{e^-} \tag{2.10}$$

$$\mathrm{Fe^{2+}} + \mathrm{Inh_{ads}} \rightarrow \left[\mathrm{Fe} - \mathrm{Inh}\right]^{2+}_{\mathrm{ads}} \tag{2.11}$$

Depending on concentration, the complex might catalyze dissolution or form protective layers.

Tantawy *et al.* proposed an inhibition mechanism for three types of inhibitors, PGS-8, PGS-10, PGS-12, of chemical structure showed in Figure 2.11, onto C1018-steel in CO_2-saturated solutions, taking into account the corrosion products of $FeCO_3$ after disassociation of CO_2 to H_2CO_3, [Tantawy, 2020, Usman, 2018]. The authors linked the efficiency of the studied inhibitors, which are cationic, to the chemical composition, including the quaternary ammonium (N^+) group, π-electrons in the aromatic-rings, and functional groups; C═O, NH, hetero O atom, and benzene ring in various directions that are responsible for functional adsorption, together with hydrophobic carbon-length of hyperconjugation. Quaternary ammonium (N^+) group adsorbs on the cathodic surface of the metal, reducing the evolution of hydrogen, and an iodide ion (I^-) adsorbs on the anodic side, reducing the anodic dissolution of the steel, in a manner that the adsorption increases with the electronic density of the inhibitor, with the hydrophobic carbon tails orienting to form a stable film that isolates the corroding surface from the solution. The schematic in Figure 2.27 represents the adsorption mechanism, driven by the chemical attraction between the unshared electron pairs of the inhibitor molecule and the vacant d-orbitals of Fe metal ions, as well as electrostatic attraction between the negatively charged metal surface (from the adsorbed negative anions of chloride and bicarbonate) and the N^+, protonated $-NH^+$ groups.

Wang *et al.* deduced adsorption mechanism of TAI from EIS measurements for the corrosion inhibition of Q235 steel [Wang, 2011]. In Figure 2.28, the 1, 2, 3 atoms in the imidazoline ring is P-π conjugated system and the ^3N atom in the imidazoline ring which is not bonding to the hydrophilic

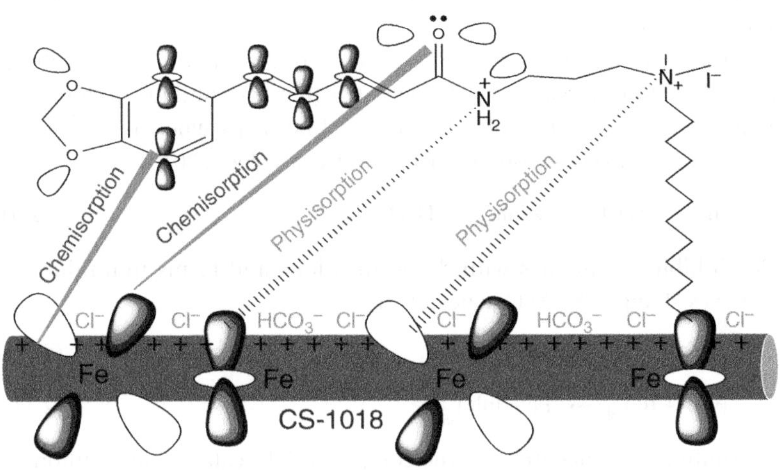

Figure 2.27 The proposed adsorption model of the synthesized surfactant inhibitor on C1018-steel in CO_2-NaCl solution. *Source:* Tantawy [2020]/with permission from Elsevier.

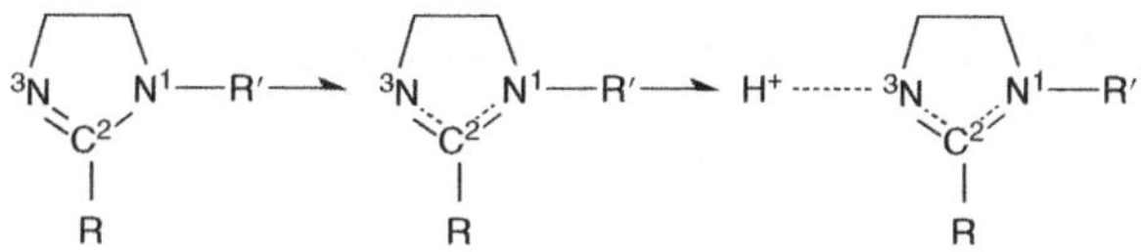

Figure 2.28 Graphical representation of protonation of imidazoline derivative in acid solution. *Source:* Wang [2011].

functional group has small steric hindrance but large electron density; so, after adding TAI to the solution, the ^{3}N atom combines with the hydrogen ion to form ammonium ion, according to [Ning, 1990], and the positively charged TAI molecules physically adsorb onto the negatively charged surface and part of the negative charges on the metal surface gets shielded, which made the potential of zero charge (PZC) shift to positive values. The authors found the shift was +60 mV in comparison to the uninhibited solution. In addition to the physical adsorption, the chemical adsorption was due to the large π-bond from the imidazoline ring and between nitrogen and sulfur atoms of the part thioureido. From that, the imidazoline ring and heteroatom are the active sites of the inhibitor, forming polycentric chemical adsorption on the surface by donating π-electron to Fe atoms and accepting electrons from 3d orbits of the Fe atoms, while the hydrocarbon chain forms a hydrophobic film, according to [Liu, 2009A].

Okafor *et al.* mentioned for the corrosion inhibition of N80 steel by 2-undecyl-1-ethylamino imidazoline (2UEI) that the inhibitor adsorbs on the surface by formation of an Fe—N coordination bond and by a π-electron interaction between the 2UEI π-electron in the head group and iron. Columbic attraction between the negative charge on the surface and the imidazoline derivative contributes also to the inhibition process [Okafor, 2009]. Lopez *et al.* proposed the corrosion mechanism, illustrated in Figure 2.29, for a commercial imidazoline-based inhibitor ($R_1 = (CH_2)_n$–$CH_3 n = 10$–17, $R_2 = (CH_2)_2$–NH_2), which has a long hydrocarbon R1 chain with hydrophobic properties, interacting to form a thermodynamically stable, bilayer molecular structure [Lopez, 2005].

Jovancicevic *et al.* proposed a similar one-dimensional model for imidazoline adsorption, in Figure 2.30, but on the corrosion product of iron oxide [Jovancicevic, 1998]. The hydrophobic group, demonstrated from linear polarization resistance (LPR) testing, had enough bilayer cohesive energy for admicelle formation. The layer had large and more lamellar micelles, and seemed to form a uniform and stable bilayer.

Liu *et al.* ascribed the inhibition effectiveness of two inhibitors AEI-11 and AQI-11 to their adsorption on the metal surface by formation of FeN coordination bond, by a π-electron interaction between the π-electron in the head

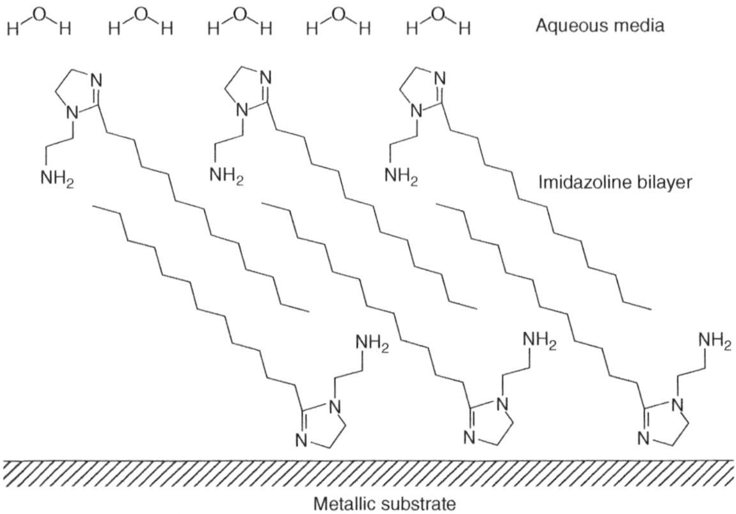

Figure 2.29 Bilayer schematic model for inhibitor commercial imidazoline-based inhibitor ($R_1 = (CH_2)_n{-}CH_{3n} = 10{-}17$, $R_2 = (CH_2)_2{-}NH_2$). *Source:* Lopez [2005].

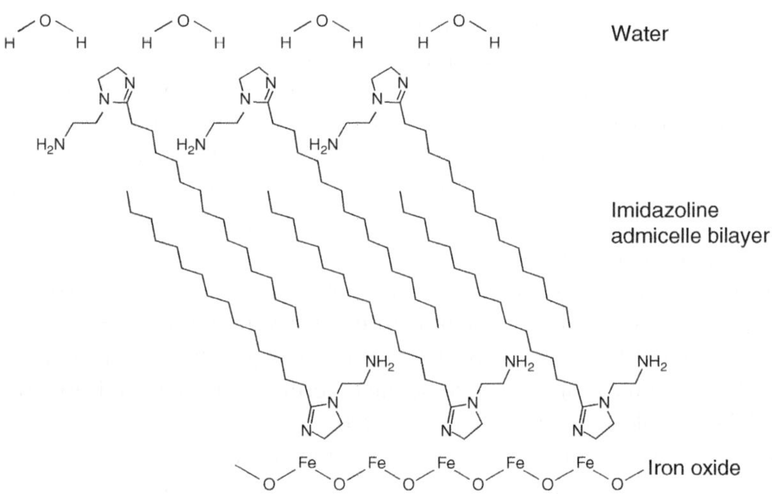

Figure 2.30 Schematic of an imidazoline admicelle bilayer on iron oxide. *Source:* Jovancicevic [1998].

group and iron, and by columbic attraction (physical adsorption) [Liu, 2009A]. Different electrochemical measurements showed that AQI-11 was more effective than AEI-11. This was because AQI-11 adsorbs on the steel surface by electrostatic interaction between the negative charge on the metal surface (from the adsorption of Cl^-) and the positive charge on the quaternary amine cation, in addition to the π-system and the formation of Fe—N coordination bond, citing also that the presence of oxygen (O_2) in the inhibitor molecule could contribute to the adsorption. On the other hand, AEI-11 adsorbs by the π-electron interaction and formation of Fe—N coordination bond.

Li *et al.* provided schematics, guided by EIS data, for the inhibition mechanisms of 1-(2-aminoethyl)-2-heptadecyl imidazoline for N80 carbon steel in CO_2-saturated 1.65 wt% NaCl solutions, depending on the concentrations of 1 and 3 g/L acetic acid (HAc), for an artificial crevice of 12 mm × 12 mm with a thickness of 300 μm [Li, 2017]. For 1 g/l HAc, the equivalent circuit and the schematic are shown in Figure 2.31. The dissolution inside the

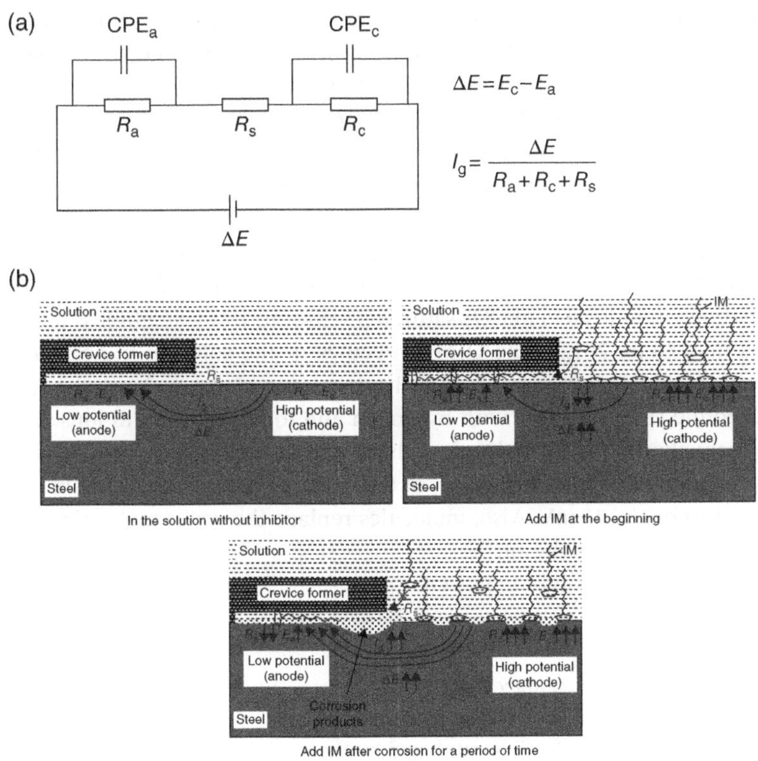

Figure 2.31 Schematic of the inhibition effect of IM inhibitor on the crevice corrosion of N80 carbon steel in the CO_2-saturated 1.65 wt% NaCl solution containing 1 g/l HAc: (a) the equivalent circuit for the galvanic effect between the electrodes outside and inside crevice, (b) IM as a mixed-type inhibitor with predominant anodic effectiveness. The number of upward arrows represents the extent of increase in parameters and the number of downward arrows represents the extent of decrease in parameters. *Source:* Li [2017]/with permission of Elsevier.

crevice was higher than outside the crevice, and the potential of the former was more negative than of the latter under the galvanic effect of the current described in the equations:

$$I_g = \frac{\Delta E}{R_a + R_c + R_s} \quad (2.12)$$

$$\Delta E = E_c - E_a \quad (2.13)$$

I_g is the galvanic current, ΔE is the potential difference between the electrodes outside and inside crevice, E_c is the potential of the electrode outside crevice, E_a is the potential of the electrode inside crevice, R_a is the reaction resistance of the electrode inside crevice, R_c is the reaction resistance of the electrode outside crevice, and R_s is solution resistance. The inhibitor, of anodic effectiveness, if added in the beginning does not enter the crevice readily due to the geometry effect, but the galvanic current decreases as the resistance in the circuit, as a whole, increases. In 3 g/l HAc solution, the inhibitor was of cathodic effectiveness, causing a significant decrease in the potential outside of the crevice, making the inhibitor of cathodic significance better for crevice corrosion.

Zhao *et al.* explained the synergistic inhibition mechanism of two inhibitors, oleic-based imidazoline (OIM) and sodium benzoate (SB), added at the same time, on mild steel in CO_2-saturated solutions from weight loss measurements [Zhao, 2012]. The OIM molecules in the solution transfer and protonate, and the nitrogen atoms, having large electron density in the imidazoline ring, combine with hydrogen and form ammonium ions. The positively charged OIM molecules adsorb physically on the cathodic sites and the point of zero charge (PZC) of the surface shift positively, estimated by the authors by 10 mV. Also, molecules replace the water molecules and adsorb chemically on the surface.

SB molecules hydrolyze to $phCOO^-$ as:

$$phCOONa \leftrightarrow phCOO^- + Na^+ \quad (2.14)$$

$$phCOO^- + H^+ \leftrightarrow phCOOH \quad (2.15)$$

Adsorption of $phCOOH^-$ is catalyzed by the positive charge of the surface induced by the OIM molecules and, with immersion time, the inhibitor film layer becomes thicker and more compact, which leads, as reported from EIS response, to higher charge transfer resistance and lower double-layer capacitance.

Xiang *et al.* analyzed the inhibition mechanism of imidazoline and piperazine on N80 steel in CO_2-saturated solutions in the presence of the impurities SO_2, NO_2, and O_2 in water phase – taking into account the reactions they undergo (in which NO_2 is involved in the formation of H_2SO_4 and HNO_3) as [Xiang, 2017]:

$$NO_2 + SO_2 + H_2O \rightarrow NO + H_2SO_4\left(\text{fast}\right) \tag{2.16}$$

$$2NO + O_2 \rightarrow 2NO_2 \tag{2.17}$$

$$2NO_2 + H_2O \rightarrow HNO_2 + HNO_3\left(\text{slow}\right) \tag{2.18}$$

$$3HNO_2 \rightarrow HNO_3 + 2NO + H_2O \tag{2.19}$$

$$4NO_2 + O_2 + 2H_2O \rightarrow 4HNO_3 \tag{2.20}$$

H_2SO_4 and HNO_3 form in the aqueous phase, and with time the following reactions occur:

$$C_4H_{10}N_2 + H_2SO_4 \rightarrow C_4H_{12}N_2SO_4 \tag{2.21}$$

$$C_4H_{10}N_2 + 2HNO_3 \rightarrow C_4H_{12}N_2\left(NO_3\right)_2 \tag{2.22}$$

In addition to the effect of neutralization, the adsorption of inhibitors by the nitrogen atoms on the surface blocks the active areas. The imino inhibitors in the system, after catalyzing neutralization reactions with H_2SO_4 and HNO_3, react with the carbonic acid or adsorb on the steel surface to block the active areas. Piperazine reacts with CO_2 as:

$$C_4H_{10}N_2 + H_2CO_3 \rightarrow C_4H_{12}N_2CO_3 \tag{2.23}$$

The authors reported that the studied inhibitors act as a mix that both stabilizes the pH and adsorbs onto the surface in supercritical CO_2 conditions containing acidic gases (which get neutralized). The inhibitors were CO_2-soluble and water-soluble.

Ortega-Toledo *et al.* reported for a commercial type HEI to adsorb on the metal surface by the formation of an Fe—N coordination bond and by a π-electron interaction between the π-electron in the head group and surface – for the molecule of a five-member ring containing nitrogen elements, a C-14 saturated hydrophobic head group and a pendant, hydrophilic hydroxyethyl group attached to one of the nitrogen atoms [Ortega-Toledo, 2010]. When the inhibitor molecules adsorb in large numbers on the surface, an electrostatic repulsion amongst the negative charges of the pendant groups occurs, causing desorption of the inhibitor molecules, and thus unprotected sites on the surface and higher corrosion rates.

Ortega-Sotelo et al. attributed the adsorption of carboxyamido imidazoline to the formation of an Fe—N coordination bond and a π-electron interaction between the π-electron in the head group and surface iron [Ortega-Sotelo, 2011]. The inhibitor consisted of a five-member ring containing nitrogen elements, a C-14 saturated hydrophobic head group, and a pendant, which is a hydrophilic carboxyamido group attached to one of the nitrogen atoms. The authors cited the columbic attraction, between the negative charge on the metal surface of electrons from the chloride ions and the imidazoline derivative as a contributor to the adsorption. When the inhibitor molecules are adsorbed in high amounts and become close enough, they desorb because of the electrostatic repulsion of the negative charges of the pendant groups.

2.3 Contact Angles of Adsorption

The phenomenon of contact angles of adsorption plays a pivotal role in understanding the interfacial dynamics between imidazoline inhibitors and the metal surface. By examining the contact angles, valuable insights can be gained regarding the adsorption behavior, wettability, and coverage of the inhibitors on the metal surface. Contact angle refers to the angle formed between the droplet of imidazoline inhibitor and the solid metal surface. A low contact angle indicates good wetting and strong adsorption of the inhibitor on the metal surface, signifying efficient coverage and enhanced corrosion protection. On the other hand, a high contact angle suggests poor wetting and limited adsorption, which may compromise the inhibitor's protective capabilities.

The contact angle of adsorption is influenced by various factors, including the chemical composition of the imidazoline inhibitor, the surface energy of the metal, and the environmental conditions. The hydrophobicity or hydrophilicity of the inhibitor molecules affects their affinity for the metal surface. In general, hydrophobic inhibitors tend to exhibit lower contact angles and better wetting, as they have stronger interactions with the hydrophobic metal surface. Conversely, hydrophilic inhibitors may show higher contact angles, indicating weaker adsorption and reduced coverage. The surface energy of the metal surface also plays a significant role in contact angle behavior. A metal surface with high surface energy tends to exhibit lower contact angles, promoting better wetting and adsorption of imidazoline inhibitors. In contrast, a metal surface with low surface energy may result in higher contact angles and reduced inhibitor coverage.

Environmental conditions, such as temperature, pH, and concentration of corrosive species, can influence the contact angles of adsorption as well. Changes in these factors can alter the interfacial properties and the behavior of imidazoline inhibitors on the metal surface. For example, an increase in temperature may enhance the mobility of the inhibitor molecules, leading to improved wetting and lower contact angles.

The measurement and analysis of contact angles of adsorption provide valuable information for optimizing the formulation and application of imidazoline inhibitors. By understanding the factors that influence the contact angles, corrosion engineers can select inhibitors with suitable wetting properties and enhance their adsorption on the metal surface. This knowledge aids in determining the optimal inhibitor concentration, surface coverage, and distribution, ensuring effective CO_2 corrosion protection in oil pipelines. Furthermore, studying the contact angles of adsorption can contribute to the development of predictive models and simulations for inhibitor behavior. These models can provide insights into the spatial distribution and arrangement of the inhibitor molecules on the metal surface, aiding in the design of efficient corrosion inhibition strategies.

Tantawy *et al.* measured the contact angles of three different cationic imidazoline inhibitors PGS-8, PGS-10, and PGS-12 on copper, aluminized paper, glass and C1018 steel substrates at 25 °C [Tantawy, 2020]. The concentration of the inhibitor droplets was 1 mg/mL and the measurements were taken with OCA 20 contact angle meter (from DataPhysics Company, Germany) after the droplets equilibrate for 15 seconds. In experiments, the inhibitors diffused from solution to the substrate/inhibitor interface with a small contact angle on copper, aluminum, carbon steel, and glass substrates. All contact angles were less than that of pure water, exhibiting good wetting capacity, and they decreased with the length of the hydrophobic chains of the PGS. PGS-12 had the smallest contact angle and therefore the highest wetting power.

Singh *et al.* evaluated the contact angle for three inhibitors – 1,3-bis(4-methoxybenzyl)-2-(4-methoxyphenyl) imidazoline (AT-1), 1,3-bis (4-methoxybenzyl)-2-phenylimidazolidine (AT-2), and 1,3-bis(4-methoxybenzyl)-2-(4-nitrophenyl) imidazoline (AT-3), added to optimum concentration [Singh, 2019]. They measured the contact angles by applying a dynamic sessile drop method on a DSA25 Drop Shape Analyzer equipped with an IDS imaging camera. The 30-μl drops without and with an inhibitor were applied onto the J55 steel specimens to determine the contact angle at 60 °C. The authors demonstrated that the J55 steel surface is hydrophilic in nature without inhibitors in an acidic solution, from a nearly 48° contact angle. With the addition of the inhibitors the wettability increased, with a protective layer that made the steel surface hydrophobic. The contact angles were nearly 115°, 94°, and 63°, respectively, for AT-1, AT-2, and AT-3.

3

Testing Setups

Testing setups are crucial for evaluating the performance of imidazoline inhibitors in mitigating CO_2-induced corrosion. Various testing methods and equipment are employed to simulate real-world conditions and assess the inhibitors' effectiveness. This section provides an overview of different testing setups commonly used in corrosion research for the sake of understanding the performance, mechanisms, and limitations of imidazoline inhibitors. They aid in the development of effective corrosion mitigation strategies, ensuring the integrity and longevity of carbon steel in oil pipeline systems.

3.1 Electrode Rotors and Stirrers

Electrode rotors and stirrers are essential components of testing setups used to evaluate the corrosion inhibition performance of imidazoline inhibitors. These devices are designed to create a controlled and uniform flow of the test solution, ensuring consistent exposure of the metal specimens to corrosive conditions. The electrode rotors facilitate the rotation of the test electrodes, promoting homogeneity of the corrosive medium and enhancing the accuracy of corrosion measurements. Stirrers, on the other hand, ensure adequate mixing of the inhibitor solution, allowing for uniform distribution and contact between the inhibitor and the metal surface. Inhibitors' performance in flowing conditions can be evaluated electrochemically in typical corrosion cells that have rotating electrodes, which are test steel samples cut and polished per corrosion testing preparation standards. Rotating electrode setups can be used to measure the influence of the shear stress on inhibitor films that form with time, or on films that form before in stagnant conditions.

Imidazoline Inhibitors for Corrosion Protection of Oil Pipeline Steels: Experimental Laboratory Evaluation and Case Studies, First Edition. Faysal Fayez Eliyan.
© 2024 John Wiley & Sons, Inc. Published 2024 by John Wiley & Sons, Inc.

The use of electrode rotors and stirrers in corrosion testing introduces potential limitations and considerations. First, the presence of mechanical components in the testing system may introduce additional sources of potential corrosion or interference. To mitigate these issues, careful material selection and design considerations are necessary. Second, while electrode rotors and stirrers provide controlled and homogeneous test conditions, they may not fully replicate the complex flow patterns and localized corrosion phenomena found in real-world applications. Therefore, researchers should consider utilizing complementary testing methods to validate findings obtained with electrode rotors and stirrers. Lastly, the design and operational parameters of electrode rotors and stirrers, including rotational speed, fluid velocity, and mixing intensity, should be carefully calibrated to match the desired testing conditions. Optimizing these parameters ensures the generation of accurate and representative corrosion data.

The corrosion test cell, similar to the one used by Yang *et al.* and Tan *et al.* to evaluate imidazoline inhibiting film stability in CO_2-saturated brines, consisted of reference electrode, gas outlet, temperature probe, platinum counter electrode, rotator, gas inlet, pH electrode, luggin capillary, working electrode and a hot plate. [Tan, 1996, Yang, 2012]. The cylindrical working electrode (WE), having a surface area between 1 and 5 cm^2, is fitted into a sample holder whose speed is adjusted with a speed controller in reference to specific pipeline flow rates. Inside the cell, a reference electrode, a counter electrode, and a gas bubbler are inserted. Common reference electrodes include saturated calomel electrodes (SCEs) and Ag/AgCl electrodes, and the counter electrodes can be made of graphite or platinum. The working, reference, and counter electrodes are connected with wires to a potentiostat that sets up the testing parameters, and controls, monitors, and analyzes the testing results.

Porcayo-Calderon *et al.* tested 1018 carbon steel as cylindrical specimens (4.55 cm long and 0.63 cm in diameter), ground by 600 grade emery paper, in the presence of 5, 10, 25, 50, and 100 ppm hydroxyethyl-imidazoline derivate inhibitor, dissolved in pure 2-propanol [Porcayo-Calderon, 2015]. The solutions were heated CO_2-saturated (3% NaCl + 10% diesel) emulsions. The cell, shown in Figure 3.1, used was prepared according to the cell described by standard ASTM G5.

The cell was a conventional three-electrode glass cell. A saturated calomel electrode (SCE, 0.242V vs. standard hydrogen electrode [SHE]) was used as the reference electrode with a Luggin capillary bridge, and a high-density graphite rod was used as the counter electrode.

In some setups, the testing cell is placed onto a magnetic stirrer that rotates a magnetic bar continuously in it to enhance with mixing the

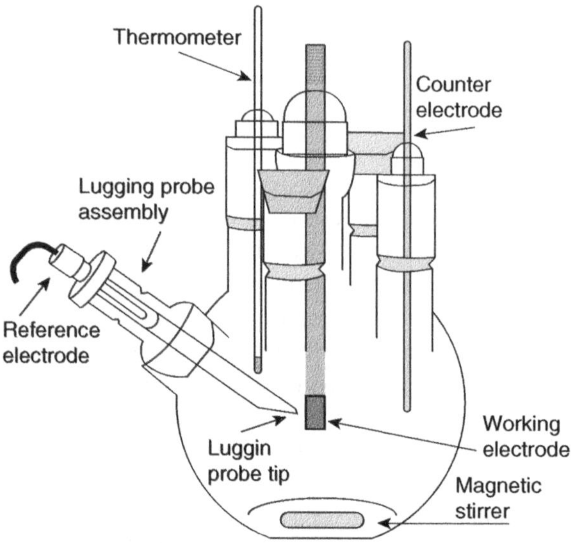

Figure 3.1 Experimental setup for electrochemical measurements. *Source:* Porcayo-Calderon [2015]/with permission of Elsevier.

homogeneity of the testing solutions. In a similar setup shown in Figure 3.2, the rotating holder has the rotating WE in its bottom part, and extra specimens fitted above it to evaluate the weight loss, immersed together in test solutions that contain brine and hydrocarbon phases [Bailey, 1997].

Ning *et al.* and Shmasa *et al.* used a glass cell setup consisting of a sparge tube, stir bar, pH probe, temperature probe, heater, condenser, and syringe. Shmasa *et al.* used the cell for testing imidazoline inhibitor on the general and localized corrosion of X65 carbon steel, in CO_2-saturated NaCl brines at 80 °C [Ning, 2013, Shamsa, 2020]. The experiments were run with and without a two-hour precorrosion period. The pre-corrosion experiments began with specimens inserted into the test brines and the inhibitor was injected after two hours. In the experiments without pre-corrosion, the specimens were inserted into the test solutions that already had the inhibitor. The total testing time was 48 hours for all experiments, and the solutions were continuously bubbled with CO_2 and mixed with a magnetic stirrer at 100 rpm to ensure mixing.

Jiang *et al.* used an apparatus consisting of a thermometer, water inlet, carbon dioxide inlet, salt bridge for the reference electrode, motor, water outlet, rotating disc electrode, counter electrode, carbon brush copper ring,

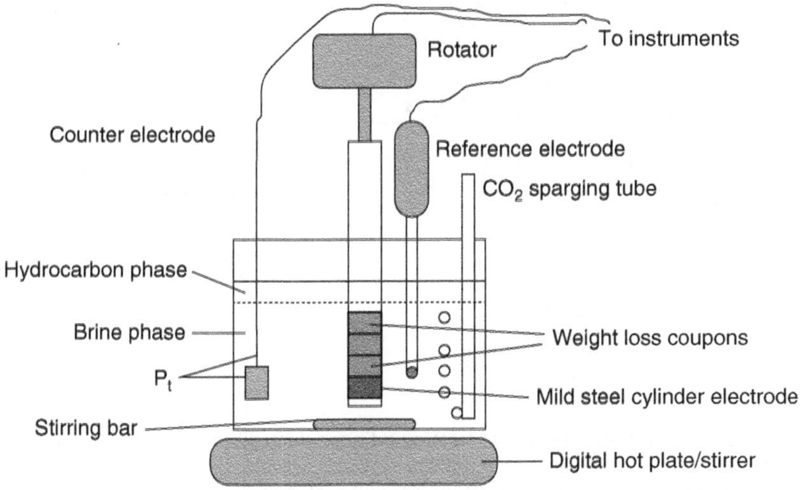

Figure 3.2 Schematic for corrosion testing setup with modified electrode rotor. *Source:* Bailey [1997]/Taylor & Francis.

baffle, and a vessel. Four samples were mounted on the rotating disk holder; one of them for electrochemical testing and the other three samples for weight loss testing, for six hours [Jiang, 2005A]. The area of a sample was $2\,cm^2$, and probes of salt bridge were symmetrically distributed and connected with each other with soft tubes and T-tube, all connected with saturated Ag/AgCl electrode to reduce the IR drop ionic electrolyte resistance between the tip of the reference electrode and the working electrode surface. Two symmetric graphite electrodes were used as the counter electrode.

Liu *et al.* evaluated the inhibition performance of three different inhibitors – thioglycolic acid (TGA), diethylenetriamine (DETA), and naphthene acid imidazolines – on bare-surface and pre-corroded N80 steel with weight loss and electrochemical methods with the same autoclave setup, with schematic shown in Figure 3.3, in CO_2-saturated 1% NaCl solutions [Liu, 2011]. Their electrode rotor HR103A was from Hokuto Denko. The shear stress between the rotating electrodes and fluid was measured from Eq. (3.1) below [Denpo, 1993]:

$$\tau_c = 0.0791 \rho u_c^2 \,\mathrm{Re}_c^{-0.3} \tag{3.1}$$

where $u_c = 2\pi r\omega$, $\mathrm{Re}_c = (r \times u_c)/\nu$, $\nu = \mu/\rho$, ω = rotating speed (rpm), r = radius of the rotating electrode (cm), ν = kinematic viscosity of test solution (cm/s), μ = dynamic viscosity of test solution (Pa s), ρ = density of test solution (g/cm^3).

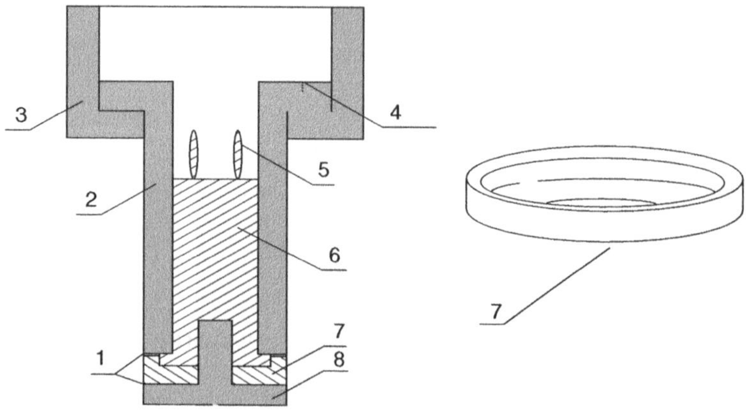

Figure 3.3 Schematic diagram of rotating cylinder electrode apparatus. (1) Sealing rings; (2) PTFE pole; (3) fixed-pressure kits; (4) fixed pinhole; (5) copper pole; (6) copper conductor; (7) working electrode ring; (8) sealing pressure sets. *Source:* Liu [2011]/John Wiley & Sons.

3.2 Concentric Cylinders in Autoclaves

Concentric cylinder setups in autoclaves are commonly employed for studying the performance of imidazoline inhibitors under high pressure and high temperature conditions. This setup consists of two concentric cylinders, where the outer cylinder acts as the autoclave vessel, capable of withstanding elevated pressures and temperatures. The inner cylinder holds the test solution containing the imidazoline inhibitor and the metal specimen. By subjecting the setup to controlled pressure and temperature, it simulates the harsh operating conditions experienced in real-world oil pipelines. This setup allows for the assessment of the inhibitory properties of imidazoline inhibitors under conditions that closely resemble those encountered in practical applications. Alternatively to rotating electrodes, which occasionally under high rotating speeds produce fluctuations in the measured potentials and currents, the test specimens can be mounted on the inner surface of an outer cylinder that is around a rotating inner Teflon cylinder, in an autoclave. A typical setup of such type is shown in Figure 3.4, which accommodates pressurized test solutions. The setup allows for measuring corrosion in both liquid and gas phases [Zhang, 2012]. The rotating cylinder moves the fluid in the clearance and allows for in-situ electrochemical measurements that are free of noise for specimens that are static in dynamic flow conditions. In the setup shown, six test specimens for weight loss and

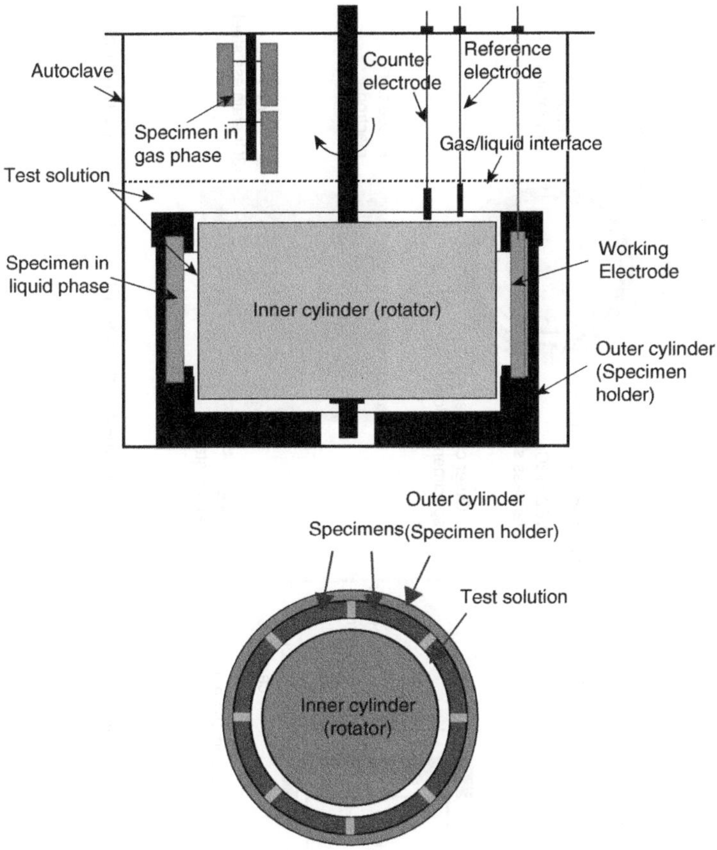

Figure 3.4 Schematic for the concentric setup in an autoclave designed for electrochemical measurements under dynamic high pressure conditions. *Source:* Zhang [2012]/with permission of Elsevier.

two specimens for electrochemical measurements were mounted on the inner side of the static cylinder in the liquid phase. In the gas phase, three specimens were mounted on the holder. The counter electrode was made of platinum and the reference electrode was Ag/AgCl. The flow velocity and shear stress on the specimens in correlation to the rotation speed of the rotating inner cylinder and its gap from the static cylinder were calculated from computational fluid dynamics (CFD) simulation.

Wang *et al.* conducted weight loss and electrochemical measurements experiments in a 3-l autoclave, consisting of double concentric Teflon cylinders, an inner cylinder rotator, and an outer cylinder holder, shown in Figure 3.5. The inner cylinder was 72 mm in diameter and the inner diameter

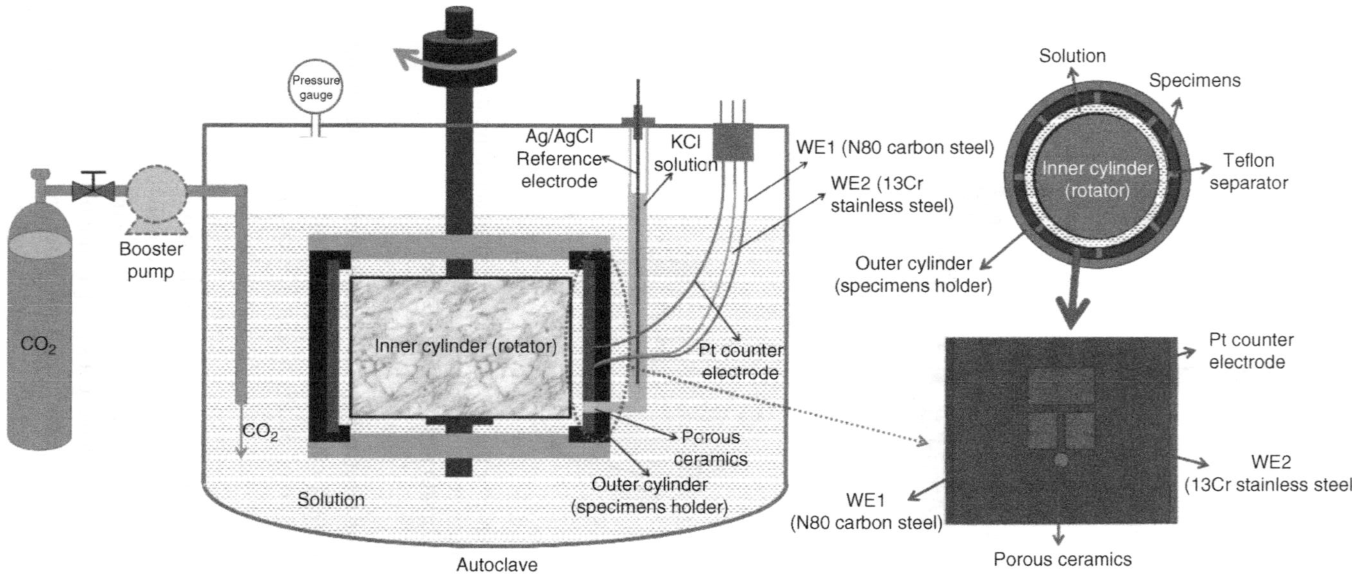

Figure 3.5 Schematic diagram of the experimental setup for in situ electrochemical measurements in supercritical CO_2-H_2O environments under dynamic conditions. *Source:* Wang [2019]/with permission of Elsevier.

of the outer cylinder was 76 mm, making a gap of 2 mm. The three-electrode system was composed of a WE, a platinum plate as the counter electrode, and an Ag/AgCl (1M KCl) electrode as the reference electrode. A carbon steel was the WE, and the platinum plate and a polytetrafluoroethylene (PTFE) tube of a porous ceramic cap were in the outer Teflon holder. The Ag/AgCl electrode was placed into the PTFE tube that was filled with 1M KCl solution). The distance between the electrodes was 0.5 mm, for reliability under high dynamic pressure flow conditions of CO_2 pressure of 8 MPa (achieved with a booster pump) and at 60 °C. The weight loss specimens were placed on the inner surface of the static outer cylinder holder against the inner cylinder rotating at 265, 530, 795 rpm, corresponding to 1, 2, and 3 m/s, respectively.

3.3 Autoclave Systems

Autoclave systems provide a versatile and controlled environment for evaluating the performance of imidazoline inhibitors. These systems consist of specially designed vessels equipped with precise pressure and temperature control mechanisms. The autoclave systems allow for the simulation of various operating conditions encountered in oil pipelines, such as high pressure, elevated temperatures, and corrosive environments. By subjecting the metal specimens to these conditions while exposing them to the imidazoline inhibitor, researchers can assess the inhibitor's effectiveness in mitigating CO_2-induced corrosion. Autoclave systems offer a valuable platform for studying inhibitor performance under controlled conditions.

Ding *et al.* used a single autoclave system for conducting experiments on the corrosion inhibition at high 120 and 150 °C [Ding, 2018]. The system contained a 4-l autoclave equipped with an inhibitor injector; three electrodes mounted through the autoclave's stainless steel lid, namely an Ag/AgCl reference probe, a Pt-coated Nb counter electrode, and the working steel electrode; and a ZrO_2-based pH probe for high temperature and high pressure conditions. The temperature in the autoclave was continuously monitored and controlled using a digital reader and the electrolyte, purged before the experiments with N_2 for deoxygenation, was stirred with an impeller at 200 rpm. In the injection system, 200 ml of 1 wt% NaCl containing the desired concentration of corrosion inhibitor was loaded into a pressure bomb, connected to the lid with appropriate fittings. Before inhibitor injection, the bomb was pressurized with N_2 so its pressure was higher than that of the autoclave; with the valve between the bomb and the autoclave opened the pressure difference pushed the inhibitor solution into the test solution in the autoclave.

The two-autoclave system has the major difference that the solutions and the specimens are heated to the desired temperatures in two different autoclaves (to limit ingress of O_2 to as low as 3 ppb, and the heating period from 3–4 hours to 20 minutes, for correspondence to the corrosion products), but the components in the single autoclave and two-autoclave systems are similar. The procedure in the two-autoclave system includes heating up and cooling down periods. In the beginning, the solution was heated at 80 °C in autoclave 2 and sparged with CO_2 continuously to remove O_2; and in autoclave 1, which contained the specimens but no electrolyte, was also sparged with CO_2 and heated to 80 °C. After two to three hours of purging, the solution was transferred from autoclave 2 to autoclave 1 with N_2 gas by opening the valve between the two autoclaves, and after complete transfer the electrochemical measurements started in autoclave 1. After the tests, autoclave 1 was pressurized with N_2 and the valve between the two autoclaves was opened for the solution to go back from autoclave 1 to autoclave 2. Cooling down of solutions and specimens were done in two separate autoclaves. Autoclave 1 was cooled to around 80 °C and depressurized before removal of the lid detached, and the hanging specimens were removed in the same procedure of the single-autoclave system. As the solutions and the specimens were cooled down separately, the cooling down period was reduced from five to six hours to two to three hours, in inert N_2 conditions for which the nature of the corrosion products did not change during the cooling period.

Chang *et al.* and Xiang *et al.* used Hastelloy C-276 autoclave by Cortest Ltd. (OH, USA) of capacity of 2.2 l to evaluate corrosion inhibition, with schematic similar to the one shown in Figure 3.6 [Chang, 1995, Xiang, 2017]. Xiang *et al.*'s simulated conditions were super critical SC-CO_2 phase (SP) and the SC CO_2-saturated aqueous phase (AP). N_2 was purged into the ionized water in the autoclave for two hours before adding inhibitors of two kinds, imidazoline and piperazine. After adjusting the sealing of the autoclave and adjusting the temperature at 60 °C, the gas mixtures CO_2/O_2 (O_2: 3.0% in mol), CO_2/NO_2 (NO_2: 5.0% in mol), and CO_2/SO_2 (SO_2: 5.0% in mol) were added in sequence into the autoclave to achieve, respectively, the pressures of 0.25, 0.40, and 0.55 MPa. Later, the high purity CO_2 was pumped with a gas booster for a high 10 MPa pressure. The gas concentrations were calculated to be 30 mol ppm, by NIST Reference Fluid Thermodynamic and Transport Properties (NIST REFPROP) software. The exhaust gas was absorbed with NaOH solution in the purifying column, and the pH of residual liquid in the autoclave was measured by a pH meter by Mettler Toledo FE20.

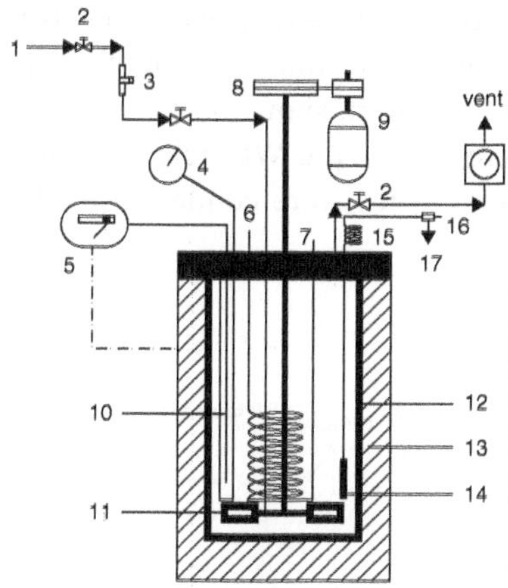

Figure 3.6 Schematic illustration of experimental apparatus for corrosion test. *Source:* Chang [1995]/John Wiley & Sons.

3.4 Crevice Testing Setup

Crevice testing setups are utilized to investigate the ability of imidazoline inhibitors to protect metal surfaces in crevice corrosion scenarios. These setups involve creating a localized crevice between two metal specimens, which simulates the conditions where crevice corrosion is likely to occur, such as gaps, joints, or sealant interfaces. The imidazoline inhibitor is introduced into the crevice and the setup is subjected to corrosive conditions. By monitoring the corrosion rate and the extent of damage within the crevice, researchers can evaluate the inhibitor's effectiveness in mitigating crevice corrosion, which can be particularly challenging to address in practical applications.

Li *et al.* used a setup of the schematic showing the configuration of the crevice for electrochemical and weight loss measurements in Figure 3.7a, b [Li, 2017]. Two identical WEs were to be soldered with copper wire as electrical connection and then sealed together with epoxy resin. One working electrode (WE1) was located inside the crevice and another one (WE2) was located outside it, of exposed area of 1 cm^2, distanced by 3 mm. Another epoxy mold was used as crevice former, and a PTFE tube (with a porous ceramic cap at one end and filled with test solution), and platinum sheet were embedded in the crevice former as salt bridge for Ag/AgCl reference electrode and counter electrode, respectively, for the electrochemical measurements inside the crevice. Another Ag/AgCl reference electrode and platinum sheet in bulk solution were used for electrochemical measurements for the electrode exposed to the bulk solution, in Figure 3.7b. A PTFE gasket with a thickness of 300 μm was located between the WEs embedded epoxy mold and the crevice former to create a crevice. The dimension of the crevice was 12 mm × 12 mm with a thickness of 300 μm, shown in Figure 3.7a. Both for the inside and outside of the crevice, as the WE, reference and counter electrodes inside and outside the crevice were connected to the

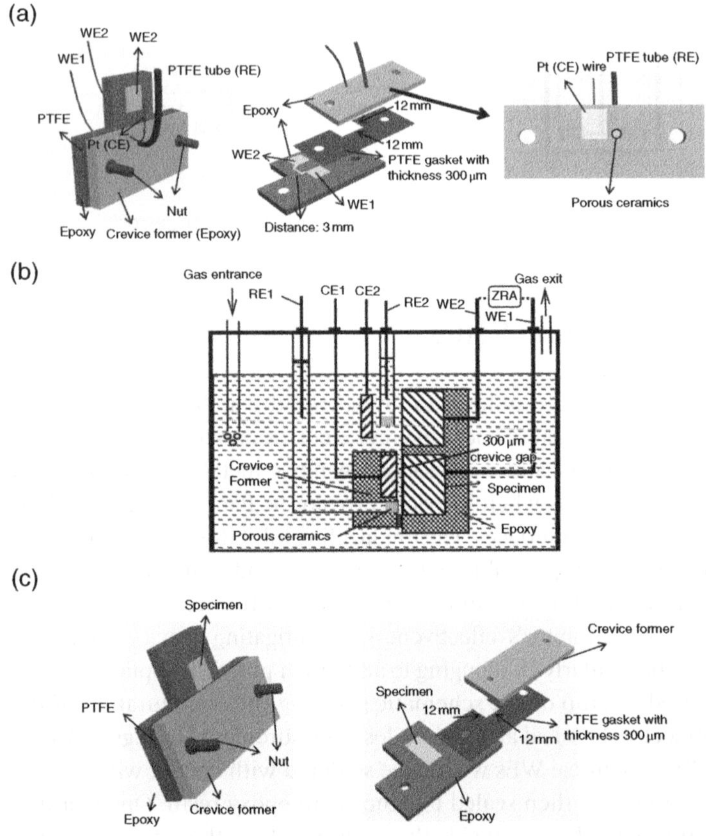

Figure 3.7 Schematic diagram of the crevice and setup for crevice corrosion: (a) configuration of crevice for electrochemical and weight loss measurements, (b) setup for electrochemical and weight loss measurements, (c) configuration of crevice for morphology analysis. (Note: When there were no electrochemical measurements, WE1 and WE2 were connected together to construct a crevice corrosion system. Electrochemical measurements were performed by temporarily disconnecting these two electrodes. WE1, CE1, and RE1 were used for the electrochemical measurements of the electrode inside crevice, while WE2, CE2 and RE2 were used for the electrochemical measurements of the electrode outside crevice). *Source:* Li [2017]/with permission of Elsevier.

workstation and the data of the WEs inside and outside of the crevice were monitored. The reference electrode was Ag/AgCl (1 M KCl). For the surface analysis, the setup in Figure 3.7c was used, for a sample of surface area of 2 cm^2, of one half inside and the other half outside the crevice, which was made in the same way as that for the electrochemical measurements.

3.5 In-situ Electrochemical/AFM Setup

In-situ electrochemical/AFM (atomic force microscopy) setups provide a powerful tool for investigating the corrosion inhibition mechanisms of imidazoline inhibitors at the nanoscale level. These setups combine the capabilities of electrochemical measurements and AFM imaging, enabling researchers to directly observe and analyze the morphological changes and interactions occurring at the metal surface in the presence of the inhibitor. By applying an electrochemical potential while simultaneously scanning the surface with AFM, researchers can obtain valuable insights into the inhibitor's adsorption behavior, film formation, and its impact on the corrosion process at the micro- and nanoscale levels.

Belarbi *et al.* conducted in-situ electrochemical/AFM measurements in a setup to confirm the adsorption of 1-(2-aminoethyl)-2-oleyl-2-imidazolinium chloride on the surface of steel in 1 wt% NaCl solutions purged with CO_2 at pH 4 [Belarbi, 2017]. The setup is Keysight 5500 Scanning Probe Microscope system held in a vibration isolation stage, equipped with n-type silicon probe was used (kn ~ 0.3 N/m, MikroMasch). The anaerobic environment was achieved by assembling the setup inside a chamber. The test cell was custom designed to have a built-in reference electrode (Ag/AgCl saturated KCl), AISI 316L stainless steel was used as the counter electrode, and the back and sides of the WE were coated with PTFE.

3.6 Flow Loops

Flow loops are sophisticated testing setups used to evaluate the performance of imidazoline inhibitors under dynamic flow conditions. These setups mimic the fluid flow encountered in oil pipelines by circulating the test solution containing the imidazoline inhibitor through a closed-loop system. The flow velocity and flow patterns can be adjusted to simulate different operational scenarios. By subjecting the metal specimens to a continuous flow of the corrosive medium while monitoring corrosion parameters, researchers can assess the inhibitor's ability to provide long-term protection under flow conditions. Flow loops offer valuable insights into the inhibitor's performance in dynamic environments, helping to ensure the integrity and reliability of oil pipeline systems.

Wang *et al.* investigated the CO_2 corrosion inhibition performance of a carbon steel in a 101.6 mm (4 in.) I.D., 15 m long Plexiglas multiphase flow loop, in which they carried out electrochemical noise measurements [Wang, 2005]. It consisted of the liquid tank, bypass pipes, liquid feed, orifice plate with mercury meter, flow height control gate, gas inlet, test compartment, gas outlet, pressure gauge, heater, valves. The test solution was ASTM saltwater with light oil LVT-200, in a 1.4 m^3 stainless steel tank (A), pumped into a PVC pipe by a 1.5 kW centrifugal pump. The flow rate was controlled by a bypass line (B) and measured using a calibrated orifice meter (D). CO_2 gas was pressurized at port F, and the multiphase mixture was passed through the Plexiglas pipeline to the tank where liquid and gas were separated using a de-entraining plate, where the liquid phase was recycled and the gas phase was exhausted at the top of the tank. The measurements were taken in the test section (G), 8 m from the gate, at which the electrochemical probes were inserted. The liquid phase was heated by two 1.5 kW heaters positioned at (K), and the temperature was controlled by a thermocouple. The slug flow was generated by controlling the gas flow at inlet (F), by a needle valve in conjunction with the gas control system. The electrochemical noise (EN) probe had three identical C-1018 carbon steel electrodes: two are WEs and the third electrode is a reference electrode.

4

Experimental Methods

4.1 Electrochemical Evaluation of Imidazolines' Performance

In the field of corrosion evaluation for imidazoline inhibitors in oil flow environments, several electrochemical techniques have been employed. Potentiodynamic polarization is a dynamic technique that provides valuable information on corrosion parameters such as corrosion potential, corrosion current density, and polarization resistance. Electrochemical impedance spectroscopy (EIS) allows for a comprehensive investigation of inhibitory behavior by analyzing charge transfer resistance, double-layer capacitance, and solution resistance. Open-circuit potential (OCP) monitoring offers insights into long-term performance and stability. Linear polarization measures the polarization resistance and evaluates inhibitory effects. Cyclic voltammetry (CV) provides information on redox behavior, adsorption processes, and stability. Electrochemical noise analysis helps assess localized corrosion suppression and overall inhibition performance. These techniques collectively offer a comprehensive understanding of the inhibitory effects of imidazoline inhibitors, enabling researchers to evaluate their corrosion inhibition capabilities and optimize their performance in oil pipeline systems.

4.1.1 Potentiodynamic Polarization

Potentiodynamic polarization is a dynamic electrochemical technique used for comprehensive corrosion evaluation of imidazoline inhibitors in oil flow environments. This technique involves scanning the applied potential linearly over a specified range, typically from a cathodic to anodic direction,

Imidazoline Inhibitors for Corrosion Protection of Oil Pipeline Steels: Experimental Laboratory Evaluation and Case Studies, First Edition. Faysal Fayez Eliyan.
© 2024 John Wiley & Sons, Inc. Published 2024 by John Wiley & Sons, Inc.

while measuring the resulting current response. By plotting the resulting polarization curve, researchers can determine important corrosion parameters such as the corrosion potential (E_{corr}), corrosion current density (i_{corr}), and the polarization resistance (R_p). The polarization curve provides valuable information about the inhibitor's ability to hinder the electrochemical reactions involved in corrosion processes and can be used to assess the inhibitory efficiency of the imidazoline inhibitor. The effectiveness of imidazoline inhibitors may vary depending on factors such as inhibitor concentration, temperature, and the presence of other corrosive species. Moreover, while potentiodynamic polarization provides valuable information about the inhibitor's performance, it might not account fully for the complex interactions and localized corrosion phenomena that may occur in real-world pipeline systems.

Potentiodynamic polarization proved vital in evaluating adsorption, stability, and desorption of imidazoline inhibitors with respect to their concentrations on pipeline steels in a variety of environmental conditions [Desimone, 2011]. The testing characterizes the inhibition performance with the conventional parameters of the corrosion potential and corrosion current density, Tafel anodic and cathodic slopes, desorption potential, and variations in current densities in the anodic and cathodic branches of the polarization at suitably chosen sweep rates [Zhang, 2001, Liu, 2009A]. Zhang *et al.* evaluated the inhibition performance of amide imidazoline-based inhibitor on iron in chloride-containing CO_2-saturated solutions with 0.2 mV/s potentiodynamic scans. They found that the inhibitor inhibited both the anodic and cathodic reactions and the corrosion potentials were more positive. With the addition of 20 ppm of the inhibitor, the corrosion efficiency was as high as 97% [Zhang, 2001]. In a different study, corrosion of API X65 steel in 5% NaCl CO_2-saturated solutions containing different concentrations of imidazoline derivative was investigated [Zhang, 2007]. The derivative inhibited both the anodic and cathodic reactions, and the kinetic parameters of corrosion potentials, current densities, anodic and cathodic Tafel slopes were extracted with the least square fit method. The corrosion potentials increased and the corrosion rates decreased with higher inhibitor concentration, with efficiency calculated at nearly 98%. The anodic and cathodic slopes of the kinetic region did not depend on inhibitor concentration. The abrupt increase in the anodic currents at high potentials indicated at desorption, as the surface coverage (θ) of the adsorbed inhibitive species decreased with the anodic potential $(d\theta/dE) < 0$, as was attributed in [Cao, 1996]. The authors mentioned that the desorption was catalyzed by the anodic dissolution under the inhibitor layer, and with the abrupt increase of the currents the surface became bare, leading to higher dissolution. Liu *et al.* evaluated the inhibition efficiency of the imidazoline derivative (1-(2-thioureidoethyl)-2-alkyl imidazoline (shown in Figure 2.13) in a CO_2-saturated brine at different temperatures [Liu, 2009B]. They found that the inhibition efficiency increased with the inhibitor concentration and temperature. The authors reported that the polarization curves shifted to more positive potentials and the anodic branches decreased with higher inhibitor concentrations, up to certain concentrations at some temperatures. From the change of Tafel slopes of the anodic and cathodic branches of the potentiodynamic polarization, whose results were in agreement with weight loss measurements at different inhibitor concentrations and temperatures, it was concluded that the tested inhibitor was of mixed type, i.e., it inhibited both the anodic and cathodic reactions. The authors estimated the interfacial capacitance of the adsorbed inhibitor film by geometric blocking from the relation:

$$y = 1 - \frac{C_c}{C_0} = \left(1 - \frac{C_s}{C_0}\right)\theta \tag{4.1}$$

where C_s and C_o are the interfacial capacitances of iron with and without the inhibitor, and C_s is the interfacial capacitance when $\theta = 1$. When η equals θ, the plot of $1 - \frac{C_s}{C_0}$ vs. θ is a straight line passing through the origin. The interfacial capacitance was extracted to be nearly $40\,\mu F/cm^2$. The inhibition at all temperatures followed Frumkin isotherm model according to the data extracted from weight loss measurements.

Singh *et al.* evaluated the interactions for J55 steel in the presence of three inhibitors - 1,3-bis(4-methoxybenzyl)-2-(4-methoxyphenyl) imidazoline (AT-1), 1,3-bis (4-methoxybenzyl)-2-phenylimidazolidine (AT-2), 1,3-bis(4-methoxybenzyl)-2-(4-nitrophenyl) imidazoline (AT-3) – at different concentrations [Singh, 2019]. With increased concentrations, the polarization curves showed similar shifts of lower current densities and less corrosion rates. This was without noticeable changes in comparison to the inhibitor-free solutions, suggesting blocking of active sites without change in the mechanisms of the anodic or cathodic reactions. The calculated Tafel slopes changed slightly, confirming that the corrosion mechanisms did not change and the inhibitor was mixed-type but of cathodic predominance as the corrosion potentials shifted to negative values with increased inhibitor concentration.

Heydari *et al.* evaluated with potentiodynamic polarization the effects of amido-imidazoline on the corrosion performance of a mild steel in CO_2-saturated solutions [Heydari, 2012]. With higher concentrations, the corrosion potentials decreased and both the anodic and cathodic currents decreased, suggesting that the tested inhibitor is of mixed type, inhibiting both the anodic and cathodic reactions. The authors noticed that the Tafel slopes did not show sensible change in relation to the absence or presence of the inhibitor in different concentrations, suggesting that the inhibition mechanism depends on the exchange current densities. From the curves, the authors concluded that the inhibition took place by geometric blocking of the active sites without getting involved in the anodic or cathodic reactions. The authors calculated the efficiency of the inhibitor with different concentrations and the corresponding surface coverage. They found that the inhibitor molecules follow Langmuir isotherm in adsorbing onto the metal surface. Desimone *et al.* studied in detail the mechanisms of *N*-[2-[(2-aminoethyl) amino] ethyl]-9-octadecenamide (AAOA) absorption from 1 mV/s potentiodynamic scans for a mild steel in CO_2-saturated 5% NaCl solutions. With an amount as low as 5.44×10^{-5} M, the corrosion rate decreased by two orders of magnitude, and further with greater concentration up to a value beyond which the corrosion rate did not show a change. At low concentrations, both the anodic and cathodic currents decreased, with no change in the corrosion potentials, suggesting that the inhibitor blocks the active sites of both the anodic and cathodic reactions, according to previous observations by Cao [1996], by physisorption. The authors attributed the change in the Tafel slopes, together with the change in corrosion potentials at higher concentrations, to chemisorption of the amine and carbonilic moieties with Fe^{2+}. The potential at which the inhibitor desorbs from the surface (E_{des}) depended on the inhibitor concentration.

Liu *et al.* investigated systematically the corrosion behavior of N80 steel in solutions containing 0.5 mmol/l of H_2S with the use of four bis(benzimidazole) derivatives containing different heteroatoms [Liu, 2012]. The inhibitors are shown in Table 4.1. The inhibitors were synthesized according to previous methods in [Matthews, 1996]. H_2S concentration was prepared with HCl (37%) and $Na_2S.9H_2O$ in the HAc/NaAc buffer solution (pH = 4.5) from which O_2 was purged. The potentiodynamic scans showed that all inhibitors reduced the corrosion current densities without Tafel slopes showing change, indicating that the inhibitors suppressed both the anodic and cathodic reactions with geometrical blocking effect. The authors tabulated the kinetic parameters. The efficiency of inhibition was in the order: BBMS > IDB > DMDB > BBP. The BBMS contains both nitrogen and sulfur and it was of best inhibition of as high as 98% with small 1 mmol/l concentration. The sulfur makes the formation of dπ–pπ bond as donor acceptor interaction of the 3d vacant orbital of Fe atom with the 3p electrons from the sulfur atom. The measurements were used to fit the surface coverage with different adsorption isotherms; with Langmuir found to be the best fit, with isotherms shown in Figure 4.1. Gibbs free energy (ΔG^{o}_{ads}) of

Table 4.1 The nomenclatures and the molecular structures of inhibitors tested by Liu *et al.* [Liu, 2012].

Nomenclature	Molecular structure	Abbreviation
1,3-Bis(benzimidazol-2-yl)propane	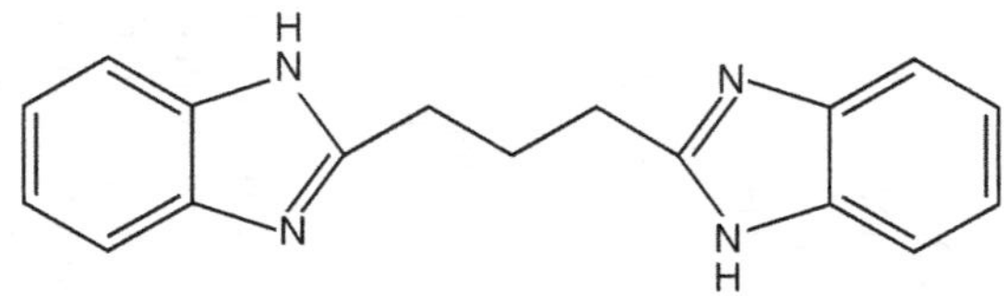	BBP
1,3-Bis(benzimidazol-2-yl)-2-oxapropane	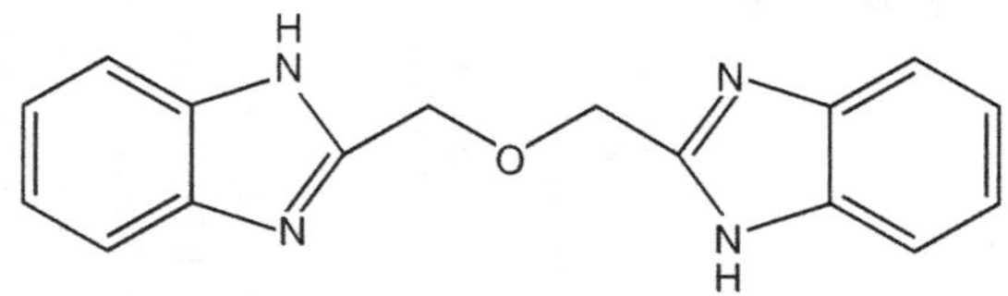	DMDB
N,N-Bis(2-benzimidazolylmethyl)amine	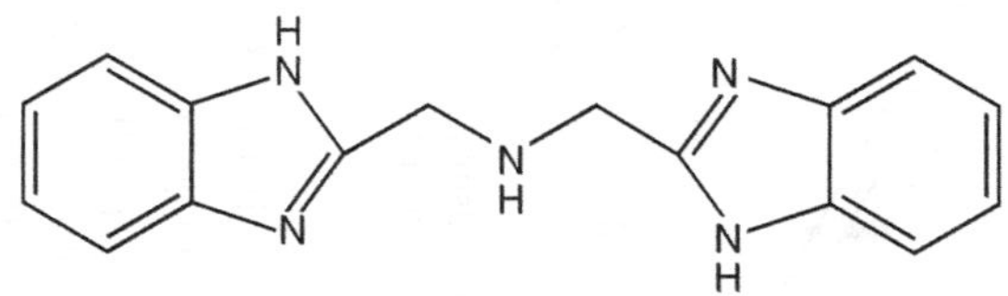	IDB
1,3-Bis(benzimidazl-2-yl)-2-thiapropane	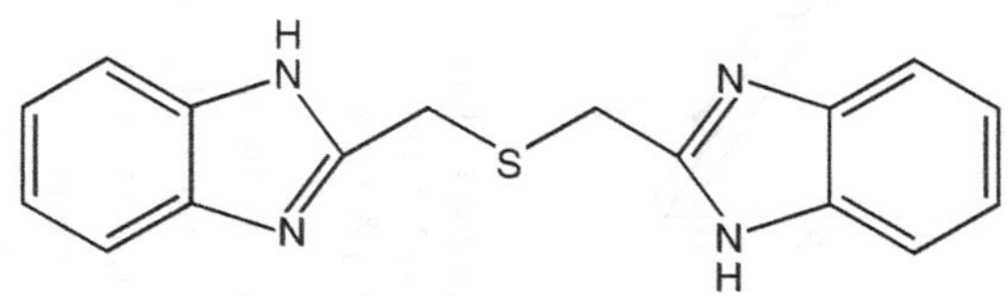	BBMS

Source: Liu [2012]/John Wiley & Sons.

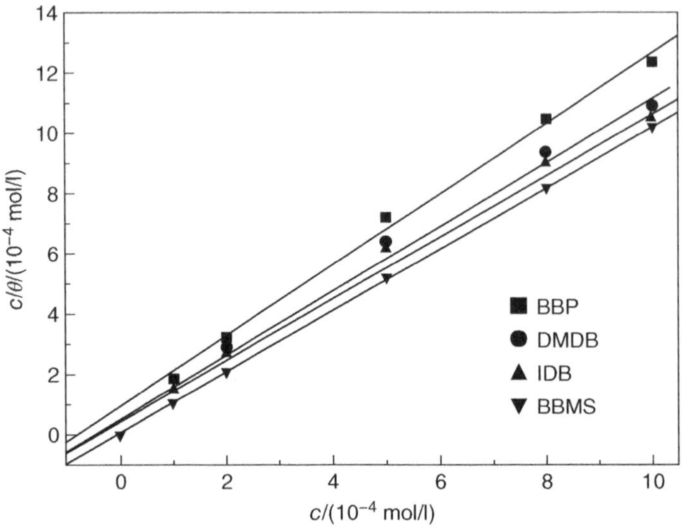

Figure 4.1 Langmuir adsorption isotherms for different inhibitors on the surface of N80 steel in H_2S solution at 20 °C, pH = 4.5. *Source:* Liu [2012]/John Wiley & Sons.

the inhibitors were about −35 kJ/mol, indicating also the spontaneous adsorption of a monolayer of the inhibitors onto the surface in both chemisorption and physisorption. The authors mentioned that the tertiary amino groups get easily protonated and the bis(benzimidazole) derivatives get easily absorbed through the interactions between the nitrogen atoms and the metal surface. The authors mentioned that the aromatic rings with π-electron cloud can jointly or individually interact with the d-orbitals of iron atoms to form covalent bonds [Muralidharan, 1995]. With metallography, the authors showed the absence of localized corrosion in the presence of the inhibitors.

Liu *et al.* tested two inhibitors 2-undecyl-1-aminoethyl imidazoline (AEI-11) and 2-undecyl-1-aminoethyl-1-hydroxyethyl quaternary imidazoline (AQI-11) on CO_2 corrosion of N80 mild steel with potentiodynamic scans of 0.166 mV/s after five hours of immersion [Liu, 2009A]. The two inhibitors inhibited both the anodic and cathodic reactions (mixed-type) and AQI-11 was more effective than AEI-11. The current plateaus in the anodic branches of the inhibited solutions were ascribed to the desorption and interaction between the inhibitor molecules and the corrosion products at nearly −0.5 V vs. SCE.

Tantawy *et al.* studied the potentiodynamic polarization response of three inhibitors of corrosion of C1018 steel in 3.5% NaCl CO_2-saturated solutions, for 50 °C [Tantawy, 2020]. The potentiodynamic scans were carried out at 0.001 V/s across a potential range of ±250 mV from the corrosion potentials. The three inhibitors, introduced in five different concentrations, were *N*-(3-((2*E*,4*E*)-5-(benzo[d][1,3]dioxol-5-yl)penta-2,4-dienamido) propyl)-*N*,*N*-dimethyloctan-1-aminium iodide (PGS-8), *N*-(3-((2*E*,4*E*)-5-(benzo[d][1,3]dioxol-5-yl)penta-2,4dienamido)propyl)-*N*,*N*-dimethyldecan-1-aminium iodide (PGS-10), and *N*-(3-((2*E*,4*E*)-5-(benzo[d][1,3]dioxol-5-yl) penta-2,4-dienamido)propyl)-*N*,*N*-dimethyldodecan-1-aminium iodide (PGS-12). Both anodic and cathodic reactions decreased in the presence of the inhibitors and there were shifts in the anodic and cathodic branches of the polarization, which were dependent on the inhibitor concentration, suggesting that the inhibitors were mixed-type. From the kinetic Tafel parameters, the inhibitors had a cathodic control and, proportionately with concentration, the corrosion current density was less, indicating a greater surface coverage accordingly.

Similarly, Zhang *et al.* investigated the effects of different concentrations of 0.488×10^{-5}, 1.22×10^{-5}, 1.22×10^{-4}, and 2.44×10^{-4} of thioureido imidazoline on X70 steel polarization in CO_2-saturated solutions [Zhang, 2019]. The samples were introduced with two different surface roughness values of 150 and 5000 grit, but the polarization, at 0.5 mV/s from 250 mV to +300 mV vs. the OCPs, did not reveal a difference. The inhibitor inhibited both the anodic and cathodic reactions, with proportional dependence on the inhibitor concentration, with desorption potentials becoming more positive accordingly. The data of the corrosion current densities and inhibition efficiencies at different temperatures corresponded to Langmuir isotherms.

Okafor *et al.* examined the corrosion inhibition of 2-undecyl-1-ethylamino imidazoline (2UEI) on N80 carbon steel in CO_2-saturated 3% NaCl solution in a controlled pH of 4 [Okafor, 2009]. The concentrations of the inhibitor were 20, 40, 80, and 100 mg/l, at 25, 40, and 50 °C. At 25 °C, two immersion times prior to the potentiodynamic polarization were tested: 30 minutes and 3 hours. The potentiodynamic sweeps were 0.2 mV/s, started after the OCPs stabilized at ±1 mV. The authors reported that the efficiency increased, the corrosion rate decreased, and the corrosion potentials became more positive proportionately with the inhibitor concentration.

The potentiodynamic profiles at 25 °C of 30 minutes of immersion indicated that the inhibitor adsorbs to block the active sites of dissolution, evident from the noticeable anodic shifts with the inhibitor concentration. The authors attributed the slight acceleration of the anodic currents with the inhibitor concentration to the corresponding higher corrosion potential. The profiles showed that the limiting current densities of the cathodic reduction of hydrogen ions decreased comparably with the inhibitor concentration, indicating the cathodic inhibition ability at a low pH 4 for corrosion reactions that were under cathodic control, and that the formation of the corrosion products was limited and slow. Increasing the time of immersion to three hours before starting the potentiodynamic scans decreased both the anodic and cathodic currents and corrosion rates. This was attributed to the formed corrosion products in combination with the adsorbed inhibitor molecules, in a different inhibition mode of geometric blocking from the active site blocking of shorter immersion period.

Zhang *et al.* studied with potentiodynamic scans in the range of −250 mV to +350 mV vs. the OCPs at a scan rate of 0.5 mV/s, the inhibitive capability of imidazoline-based inhibitor made from oleic and diethylenetriamine in CO_2-saturated solutions at 60 °C [Zhang, 2018]. With higher concentration, the corrosion potentials were more positive, the corrosion rates significantly decreased, and the inhibition efficiency was as high as 98%, with indications that the inhibitor affected both the anodic and cathodic reactions.

The authors utilized differential polarization curves to evaluate the adsorption stability of the inhibitor for two test steels. The *Y*-axis represented a semi-logarithmic slope of anodic polarization curve, defined as [Tao, 2012]:

$$D = \frac{d \log i}{\mathrm{d}E} \tag{4.2}$$

With the increase in the concentration, the potentials of desorption shifted to higher values for both steels, indicating a higher stability and inhibitor adsorption correspondingly.

Wang *et al.* used potentiodynamic polarization to study the significance of galvanic coupling of N80 steel with 13Cr stainless steel in the presence of imidazoline (IM), 1-butyl-2-thiobenzylbenzimidazole (BTBBI) and lauric acid (LA), with concentrations of 100 mg/l, in dynamic supercritical CO_2 oilfield water conditions [Wang, 2019]. With the addition of inhibitors, the anodic and cathodic current densities and corrosion current densities decreased, and the corrosion reactions were activation-controlled. From the values of corrosion potentials, the inhibitors were mixed-type, but of anodic significance. The polarization data of 13Cr stainless steel suggested that the inhibitors could have interfered with the protectiveness of the corrosion products as the pitting potentials were lower.

Khavasfar *et al.* carried out potentiodynamic polarization scans in relation to concentrations of commercial-type imidazoline, for a mild carbon steel immersed in CO_2-saturated solutions containing 96.2 g/l NaCl, 3.3 g/l $CaCl_2$ and 1.8 g/l $MgCl_2$, proposed by NACE, as NACE1D-182 simulating gas well solution [Khavasfar, 2007]. With higher inhibitor concentration from 100 to 500 ppm, the corrosion rates decrease and the corrosion potentials get higher.

Okafor *et al.* evaluated the corrosion performance of N80 carbon steel in CO_2-saturated simulation formation solutions in the presence of 50, 100, and 200 ppm rosin amide imidazoline (RAIM), with 0.2 mV/s polarization scans [Okafor, 2011]. The inhibitor changed the cathodic reactions to be more under charge-transfer control than being under diffusion control in the absence of the inhibitor – during the nonhomogeneous reaction of CO_2 hydration, which gets inhibited proportionally with inhibitor concentration. The anodic reactions in the presence of the inhibitor exhibited two linear anodic slopes, and both anodic and cathodic currents were less positive and the corrosion potentials were more positive, indicating that the inhibitor was of mixed-type, blocking the active sites.

Migahed *et al.* evaluated the inhibition performance of 1-(2-aminoethy l)-1-dodecyl-2-hexadecyl-4,5-dihydro-1H-imidazol-1-ium (PQI) and 1-(2-aminoethyl)-1-dodecyl-2-heptadecyl-4,5-dihydro-1H-imidazol-1-ium (SQI) in different concentrations from 50 to 500 ppm in oil well water formation solution for X-65 carbon steel, with potentiodynamic scans shown in Figure 4.2 [Migahed, 2017]. The scans were in the range of 300 mV to −300 mV, relative to the OCPs, at a rate of 2 mV/s. Both inhibitors were of mixed type as the greatest change in corrosion potentials, relative to the corrosion potentials of blank solutions, did not exceed 20 mV. The inhibitors decreased the corrosion current densities and anodic and cathodic currents proportionally with the inhibitor concentration. The authors attributed the no change in Tafel anodic and cathodic slopes with the different inhibitor concentrations to the independence of the charge-transfer mechanisms from the inhibitor upon adsorption (as geometric blocking).

The authors, with consideration of the interaction energies of PQI and SQI inhibitors with the carbon steel surface that were higher than those of water molecules, constructed adsorption isotherms. The surface coverage (θ) obtained from the polarization data showed fitting with Langmuir adsorption isotherms of straight lines approaching unity in slopes (attributed to shared repulsion or attraction in the interaction between the inhibitor molecules and the surface). The lines, of curve regression coefficients approaching unity, satisfied the equation below:

$$\frac{C_i}{\theta} = \frac{1}{K_{ads}} + C_i \tag{4.3}$$

where C_i and K_{ads} are the molar concentration of the inhibitor and the standard adsorption constant, respectively. K_{ads} values indicated at strong adsorption, and ΔG values were negative, nearly −14 kJ/mol, indicating spontaneous physisorption.

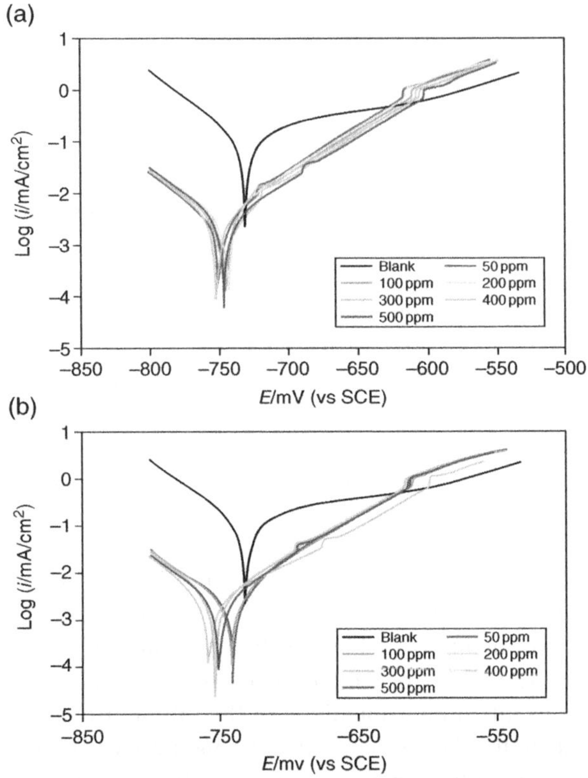

Figure 4.2 Potentiodynamic polarization curves for carbon steel in formation water in the absence and presence of various concentrations of (a) PQI inhibitor and (b) SQI inhibitor. *Source:* Migahed [2017]/Springer Nature.

Ortega-Toledo *et al.* experimented a commercial type hydroxyethyl imidazoline (HEI), dissolved in pure 2-propanol, in CO_2-saturated 3% NaCl solutions heated at 50 °C. The inhibitor was introduced after 30 minutes of exposure, in 5, 10, 25, 50 and 100 ppm [Ortega-Toledo, 2010]. The polarization scans were from −0.3 to 0.5 V vs. the OCP, at a rate of 1 mV/s. With higher inhibitor concentration, the corrosion potentials shifted to higher values and the corrosion current densities were less accordingly, attributed to active site blocking effect. The highest inhibition efficiency was nearly 60% attained at 25 ppm. At high anodic potentials, the currents even in the presence of high inhibitor concentrations were high, and the cathodic currents were not affected by the inhibitor, indicating that the corrosion was under cathodic control.

With the use of 20 ppm of organic HEI inhibitor, Rivera-Grau *et al.* monitored the change in corrosion potentials with time, for a 1018 carbon steel,

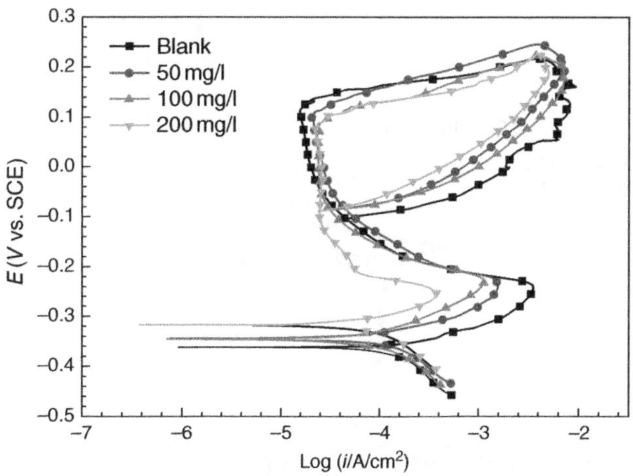

Figure 4.3 Potentiodynamic polarization curves of N80 steel in the CO_2-saturated NaCl (0.2 mol/l) + $NaNO_2$ (0.2 mol/l) solutions containing different concentrations of TAI. *Source:* Zhao [2018]/Elsevier/CC BY 4.0.

in 3% NaCl + diesel CO_2-saturated solution at 50 °C [Rivera-Grau, 2016]. For 24 hours, the potentials showed slow change, being nearly −0.73 V vs. SCE, during 24 hours of immersion.

Zhao *et al.* used 1-(2-thioureidoethyl)-2-alkylimidazoline (TAI), added in 50, 100, and 200 mg/l to inhibit the corrosion of N80 steel in CO_2-saturated 0.2 mol/l NaCl + 0.2 mol/l $NaNO_2$ solutions (pH ~ 5), in cyclic potentiodynamic polarization tests, of curves shown in Figure 4.3 [Zhao, 2018]. Before the experiments, for 30 minutes of immersion, the variations in the OCPs were nearly 5 mV. The authors reported that the scan rate had no noticeable influence on shape of the curves for all tests, which were repeated three times, starting at −0.1 V vs. OCP, at a chosen scan rate of 0.5 mV/s. The corrosion potentials were more positive with the inhibitor concentration, indicating that the inhibitor is of anodic type. The inhibitor enhanced the passivation process, but also the pitting process, as the passive currents were less but the pitting potentials were lower. This was attributed to adsorption of the inhibitor in the beginning and to the desorption that cause defects in the passive film at higher potentials, onto which the chloride ions adsorb preferentially to drive the pitting process.

Ortega-Sotelo *et al.* used commercial carboxyamido imidazoline, introduced in 1.6×10^{-5}, 3.32×10^{-5}, 8.1×10^{-5}, 1.6×10^{-4}, and 3.32×10^{-4} M, in CO_2-saturated 3% NaCl solutions at 50 °C [Ortega-Sotelo, 2011]. The inhibitor was added after two hours of X70 steel specimen pre-corrosion, in measurements carried out one hour later. The scan rate was 1 mV/s, from −0.3 to

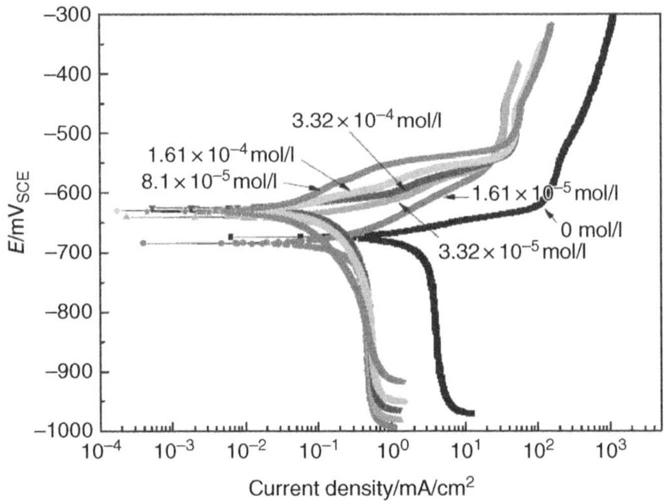

Figure 4.4 Effect of carboxyamido imidazoline concentration in the polarization curves for X-70 pipeline steel in CO_2-saturated 3% NaCl solutions.
Source: Ortega-Sotelo [2011]/Springer Nature.

0.3 V vs. SCE, vs. OCP, for the curves shown in Figure 4.4. The anodic currents decreased with the inhibitor concentration to 8.1×10^{-5} M, to increase again to 3.32×10^{-4} M. The cathodic currents were not as sensitive to the inhibitor, indicating that the corrosion was controlled by the cathodic reduction and the inhibitor was anodic-type. The corrosion potentials and corrosion current densities increased and decreased, respectively, with the inhibitor concentration up to a certain concentration. At potentials below −0.55 V vs. SCE, the authors attributed abrupt increases in the anodic currents to the desorption of the inhibitor molecules of unstable films at high potentials and high dissolution rates of the surface. The authors cited the cathodic reduction of the carbonic acid and bicarbonate species, and the precipitation of iron carbonate during anodic dissolution.

4.1.2 Electrochemical Impedance Spectroscopy

Electrochemical impedance spectroscopy (EIS) is a powerful technique for investigating the corrosion inhibition properties of imidazoline inhibitors in oil flow environments. EIS involves applying an AC signal to the working electrode over a wide frequency range and measuring the resulting impedance response. By analyzing the complex impedance spectra, researchers can extract important electrochemical parameters such as charge transfer resistance (R_{ct}), double-layer capacitance (C_{dl}), and solution resistance (R_s).

These parameters provide insights into the inhibitive behavior of the imidazoline inhibitor, including its ability to form protective films, hinder charge transfer processes, and modify the electrochemical reactions occurring at the metal surface. The interpretation of impedance spectra can be complex, and the presence of adsorbed species, electrode roughness, and other factors can influence the impedance response. In many of the experimental evaluations, frequency range and fitting models used in EIS analysis can impact the accuracy of the results.

The use of EIS proved effective in evaluating, from measuring the interface capacity and other parameters to the adsorption and desorption of imidazolines, in a variety of environmental conditions set up in laboratory [Tan, 1996]. Heydari *et al.* investigated the effect of different concentrations of amido-imidazoline on the inhibition performance of API 5L X52 during 30 minutes of immersion in 3% NaCl CO_2-saturated solutions at 25 °C [Heydari, 2012]. During short periods of immersion, the behavior in the absence of the inhibitor was highly capacitive and was apparently controlled by charge transfer [Keddam, 1981]. Increasing the inhibitor concentration changed the behavior and increased the size of the capacitive arcs, and the nature of the interactions depended on the inhibitor concentration. Across the inhibitor film, the interactions were diffusion-controlled, and its resistance from the equivalent circuit calculations was proportional to the inhibitor concentration and it remained in value less than the charge-transfer resistance. Desimone *et al.* found that the addition of 27.2 μmol/dm^3 of AAOA made the interfacial interactions in CO_2-saturated chloride solutions dependent on temperature, unlike with the uninhibited solutions [Desimone, 2011]. The authors proposed different electrochemical equivalent circuits that had good fit accordingly with different conditions.

Tan *et al.* evaluated in CO_2-saturated 3% NaCl solutions an imidazoline, added in 50 ppm, film stability and development with time, across which EIS response was regularly measured [Tan, 1996]. The Nyquist EIS spectra for mild steel of semicircles were getting larger with time, suggesting a continuous increase of imidazoline adsorption and gradual interference with the kinetics of the surface. For each of the Nyquist plots, two capacitive, overlapped semicircles appeared. The low-frequency region corresponds to the interfacial electrochemical process of charge transfer and the high-frequency region corresponds to the inhibitor film. The inhibitor film, in a specific time constant, was of a specific phase shift that increased with time. The authors suggested that with time the inhibitor film grew in layers of different physicochemical properties that each were modeled with a specific time constant, in an equivalent circuit to which the experimental data were fitted, with indication at the inhibitor layer. The inner, first layer was suggested to be comprised of inhibitor metal complexes and the outer layers had successively less molecular density across which the inhibitor molecules were cross-linked. The authors also evaluated the inhibitor destruction by the shear stress of the fluid in which the cylindrical sample was rotated after the inhibitor film formed. The Nyquist spectra decreased in size with time and the changes in the interface were reflected in the calculated values of the surface and inhibitor capacitances.

Similarly, recently, an imidazoline film in CO_2-saturated solutions was modeled with an equivalent circuit that accounted for different layers in solutions in which the effects of imidazoline concentration, temperature, and stirring were evaluated [Doležal, 2013]. The equivalent circuit together with the Nyquist spectra are shown in Figure 4.5. Zheng *et al.* in an early

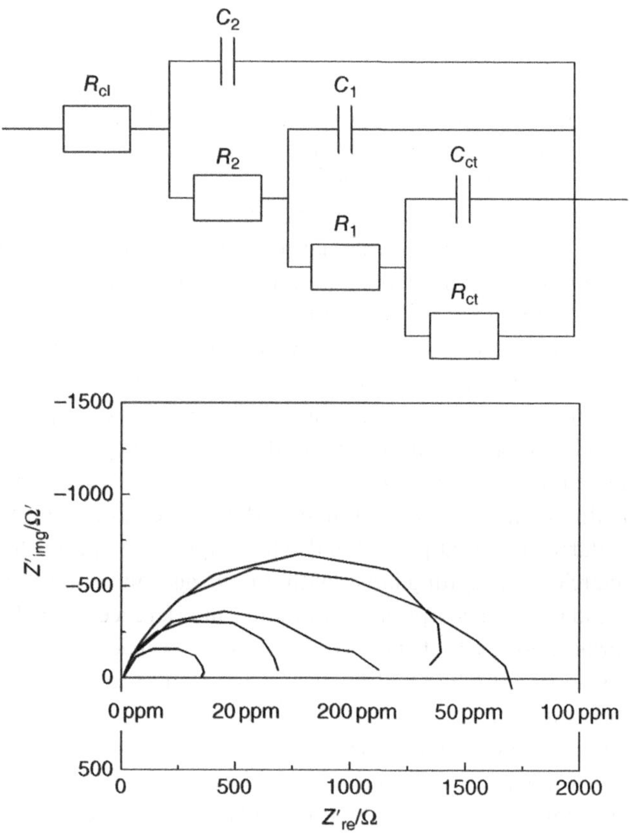

Figure 4.5 The equivalent circuit and the Nyquist spectra representing different imidazoline concentrations after 22 hour of immersion of carbon steel in CO_2-saturated brines. *Source:* Doležal [2013]/John Wiley & Sons.

study linked typical EIS responses for amide imidazoline to the inhibitor concentration and the corrosion mechanisms [Zhang, 2001]. From the equivalent circuits of different solutions, the interface capacity was used to calculate the efficiency of inhibition.

Imidazoline together with polymerized vegetable fatty acids was used as a filming fluid for EIS measurements for a mild steel in CO_2-saturated NaCl solutions. The filming fluid consisted of the inhibitor added in 16 ml and brine-diesel emulsion (640 ml of 3% NaCl brine and 160 ml of diesel) [Bailey, 1997]. The inhibitor film formation, deterioration, and persistency were studied with EIS. The Bode and Bode phase of the EIS response in the fluid is shown in Figure 4.6, for the time period of 10, 20, and 30 minutes,

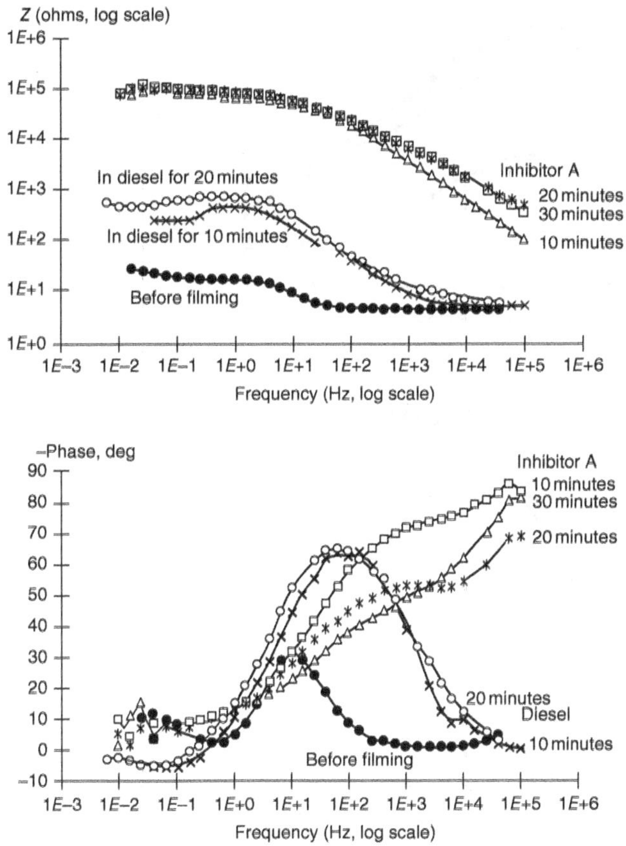

Figure 4.6 Bode and bode phase of the impedance response for mild steel in filming fluid containing 16 ml of imidazoline inhibitor in CO_2-saturated solution. Bailey [1997]/Taylor & Francis.

and those for diesel-filmed electrode and bare electrode. The low-frequency impedance of the inhibitor-filmed electrode was apparently notably larger than the impedance of the diesel-filmed and bare-filmed electrodes. The Bode phase response of the inhibitor-filmed electrode was quite different from that of the diesel-filmed electrode.

The inhibitor-filmed electrodes were exposed to 3% NaCl solutions under stagnant conditions for 2, 14, and 17 hours, showing changes that reflect distinctively the quality of the inhibitor film with time, as shown in Figure 4.7. The large high-frequency peak became with time two small new peaks and the electrode impedance decreased, indicating the gradual

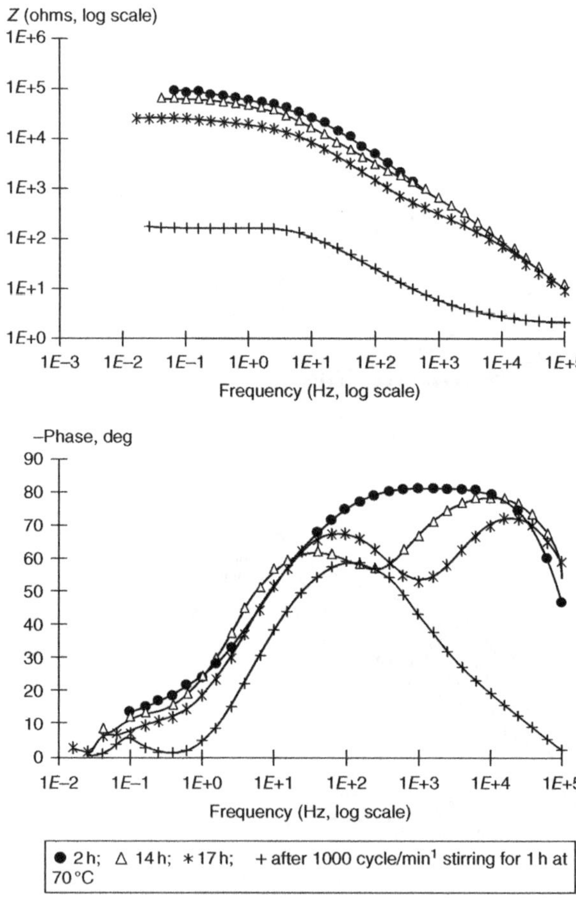

Figure 4.7 Bode and bode phase of the impedance response for mild steel after exposing it to 3% NaCl after being filmed in filming fluid containing 16 ml of imidazoline inhibitor in CO_2-saturated solution. *Source:* Bailey [1997]/Taylor & Francis.

deterioration of the inhibitor film. In one condition, the electrode was rotated at 1000 cycle/min for one hour at 70 °C. The Bode spectra clearly indicated the total destruction of the inhibitor film, as the high-frequency phase shifts disappeared. Under shear stress and high temperature, imidazoline inhibitors showed poor persistency. The authors investigated the corrosion rate during film deterioration of data of the double layer superimposed with that of the inhibitor film. They related the weight loss observations to the changes of the bode spectra. In addition, together with the fitting of the Nyquist spectra and calculation of the electrical elements of the equivalent circuits, the deterioration of the inhibitor film was empirically analyzed, attributed to the inhibitor film porosity that was accounted for with a resistance element in the equivalent circuits.

Zhang *et al.* investigated with EIS the interfacial changes during 72 hours in the presence of 50 ppm of an imidazoline derivative in CO_2-saturated 5% NaCl brine [Zhang, 2007]. The Nyquist and Bode impedance response were used to describe the interactions and the adsorption of the inhibitor in a bilayer. The continuous increase in the size of the Nyquist plots was attributed to the continuous coverage of the surface. In the Bode phase plots, a second time constant appeared during 24 hours of immersion. The one-time constant at a frequency of 1 Hz and the other at 1800 Hz were respectively attributed to the double layer and a relaxation process of the adsorbed species. The phase shift and appearance of a new phase with time indicated that the inhibitor was adsorbed in a bilayer structure.

Desimone *et al.* examined the interfacial changes during two hours of immersion of steel in CO_2-saturated solutions with EIS, in reference to the concentration of AAOA [Desimone, 2011]. In their analysis, they attributed part of the capacitive behavior in the Nyquist response to the roughness and inhomogeneities of the surface. The Nyquist arcs increased in size, in agreement with polarization and OCP measurements with higher inhibitor concentration, indicating lower corrosion rates. The equivalent circuits and the number of the time constants changed with the inhibitor concentration. The values of phase angles increased with the inhibitor concentration, corresponding, together with the double layer capacitance, to porous inhibitor films blocking the active anodic sites. The authors attributed the low-frequency time constants to corrosion interactions taking place at locations where the solution penetrated through the porous inhibitor film. The high-frequency time constants, with higher inhibitor concentrations, were attributed to stable and protective inhibitor films. The authors indicated the mechanism of the inhibitor adsorption as chemisorption of the organic elements in the inhibitor. This was suggested from the steady drop of the frequency of the maximum of the imaginary component of the impedance.

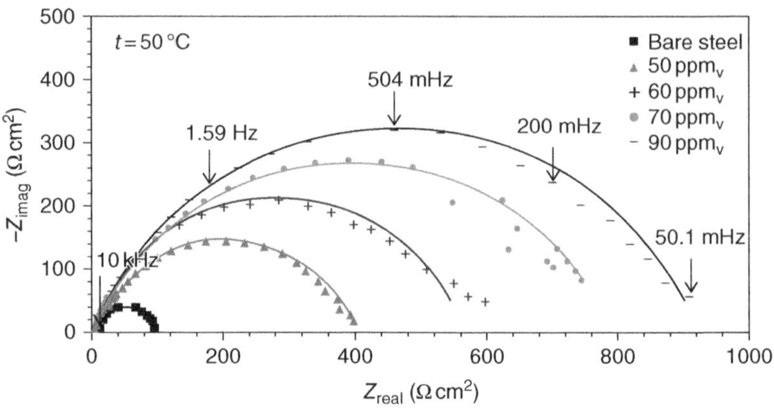

Figure 4.8 Nyquist plots for a mild steel in CO_2-saturated 3 wt% NaCl solution with and without different concentrations of TOFA/DETA imidazoline at 50 °C. *Source:* Jevremović [2016]/John Wiley & Sons.

Jevremović *et al.* analyzed the adsorption of tall oil diethylenetriamine imidazoline (TOFA/DETA imidazoline) on a mild steel in CO_2-saturated 3 wt% NaCl brine at different temperatures with EIS [Jevremović, 2016]. Regardless of temperature, the mechanisms of corrosion and the nature of inhibition did not change. Figure 4.8 shows Nyquist plots of highly capacitive interfaces of the adsorbed inhibitor that did not depend on the concentration, different from other imidazoline inhibitors that with concentration could change the interfacial interactions of the inhibition accordingly.

The authors extracted from the equivalent circuit the charge-transfer resistance and used them to describe the inhibitor surface coverage and to construct Langmuir adsorption isotherms, shown in Figure 4.9. The authors attributed the slight nonlinearity in the isotherms to postulated interactions between the adsorbed and adsorbing inhibitor molecules, in addition to the size of the hydrophobic chain in TOFA/DETA imidazoline molecule that result in steric hindrance for adsorbing molecules. Temkin adsorption isotherms, which account for the molecules interactions, were constructed. The authors calculated the thermodynamic parameters associated with the adsorption, Gibbs free energy, enthalpy, and entropy, from the Langmuir and Temkin adsorption isotherms, and mentioned comparatively with other studies that it was difficult to determine the adsorption mechanism (physisorption or chemisorption) of the inhibitor based on the calculated parameters. They indicated the cationic nature of the adsorption between the positively charged inhibitor molecules and the negatively charged metal surface.

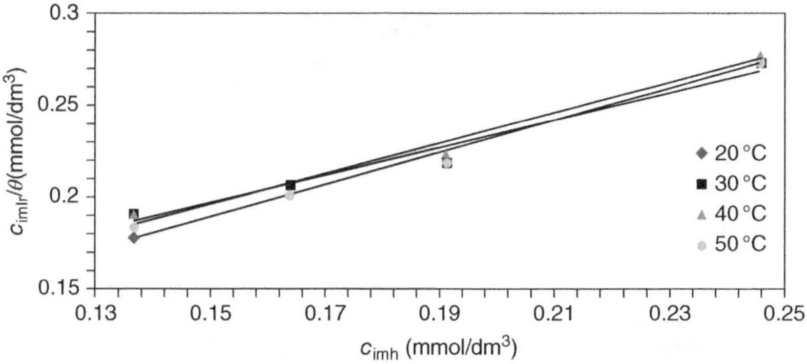

Figure 4.9 Langmuir adsorption isotherms for mild steel in CO_2-saturated 3 wt% NaCl solution in the presence of TOFA/DETA imidazoline at different temperatures. *Source:* Jevremović [2016]/John Wiley & Sons.

Chen *et al.* studied the synergistic effect of the oxide film and imidazoline in protecting the surface of L360 steel in CO_2/H_2S environments with EIS and proposed equivalent circuits that fit the interactions accordingly with different inhibitor concentrations [Chen, 2015]. The experiments were carried out at 60 °C in solutions that contained 1 : 1 CO_2/H_2S. The solutions contained anions that simulate the stratum water that runs in the multiphase in oil pipelines, including potassium, sodium, chloride, ammonium, sulfate, magnesium, and calcium. Expectedly, the corrosion resistance was higher with greater inhibitor concentration showing capacitive loops that increased in size and without indication at change in the protection mechanisms, both for the bare surfaces and surfaces covered with oxide films, as shown in Figure 4.10.

The authors reported the surface covered with oxide film with inhibitor as complicated to model and they utilized the scanning electron microscopy to help analyze the EIS results. The oxide film was compact, but in some areas, there were pores through which imidazoline diffused, increasing the layer compactness and in turn the charge transfer resistance. With higher concentrations, the inhibition efficiency was higher and the inhibitor adsorbed onto the oxide film. The authors modeled the interactions across the interface. Tantawy *et al.* demonstrated that the capacitive loops increased in size with concentration [Tantawy, 2020]. The authors reported high capacitive capability of three inhibitors in saline CO_2-saturated solutions at 50 °C, in comparison to the interfacial EIS response of the inhibitor-free solution. The authors attributed the shape of the capacitive loops to the roughness of the test C1018 steel surface and inhomogeneities, for which the constant phase element (CPE) was added for the equivalent

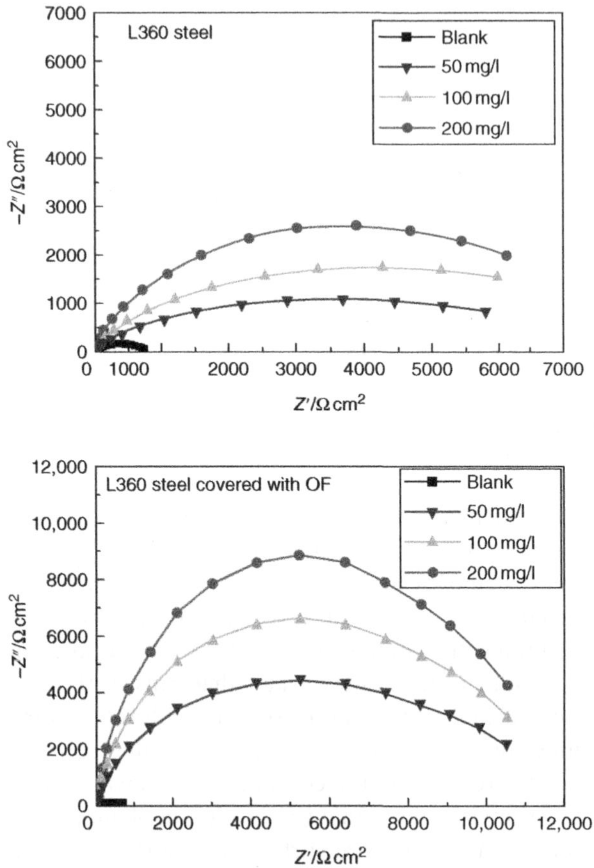

Figure 4.10 EIS plots from a WE with and without OF in the CO_2/H_2S saturated solution in the presence and absence of IM at 60 °C. *Source:* Chen [2015]/Emerald Group Publishing Limited.

circuit of the inhibited surfaces in addition to the inhibitor adsorption. From the Bode and Bode phase diagrams, the authors deduced that metal/solution interface approaches the response of an ideal capacitor at a higher inhibitor concentration, at which tested inhibitor formed a pseudo capacitive protective layer.

Liu *et al.* evaluated the performance of two inhibitors AEI-11 and AQI-11 on CO_2 corrosion of N80 mild steel with EIS of 100 kHz–10 MHz with a signal amplitude perturbation of 5 mV at corrosion potentials [Liu, 2009A]. The Nyquist spectrum of AQI-11 was noticeably larger than AEI-11, and the charge transfer resistance correspondingly was larger, indicating higher efficiency and confirming the potentiodynamic results. Wang *et al.* studied the inhibition performance of 1-(2-thioureidoethyl)-2-alkylimidazoline inhibitor (TAI), added in 0.08, 0.15, 0.23, and 0.31 mmol/dm^3, on the corrosion of Q235 steel in an aqueous solutions of 2% NaCl saturated with CO_2, during different immersion times at 293 K (20 °C), depicted with Nyquist plots [Wang, 2011]. With increased concentration, the loops increased in size from high to medium frequencies, with inductive loops at low frequencies indicating the adsorption of intermediate corrosion products from anodic dissolution in acidic/CO_2-saturated environments, proposed by other researchers [Nesic, 1996, Bockris, 1961]. From the inductive loops, the authors indicated the significance of diffusion in the presence of the inhibitor, and the depressed semicircle shapes indicated the inhomogeneities of the test surfaces, as reported before by other researchers such as in [Lopez, 2005]. Accordingly, different equivalent circuits that included equivalent phase elements were made to fit the interfacial EIS response. From analyzing the charge transfer and capacitance of the interfacial layers, the authors concluded that when the inhibitor concentration was lower than 0.15 mmol/dm^3, the inhibitor adsorbed as molecules parallel to the surface including the long hydrocarbon chain, decreasing the active sites and repelling water molecules, and the greater the inhibitor concentration, the greater the surface area was covered. However, beyond 0.15 mmol/dm^3, the inhibitor molecules tended to electrostatically repulse each other, making the long hydrocarbon chains to adsorb perpendicularly on the surface, covering fewer active sites and causing a drop in the inhibitor efficiency. The perpendicular adsorption was proposed in similar investigations that described the change in the inhibition mode and the corresponding decrease in inhibition efficiency with higher inhibitor concentration [Jiang, 2005C, Singh, 1993].

Okafor *et al.* used EIS to evaluate the corrosion inhibition performance of 2UEI on N80 mild steel in CO_2-saturated 3% NaCl solutions in different conditions [Okafor, 2009]. All tests were carried out at controlled pH 4 conditions, and the concentrations of the inhibitor were 20, 40, 80, and 100 mg/l, with consideration of the immersion times before starting EIS tests of 30 minutes and 3 hours, when free potentials stabilized at ±1 mV over frequency range of 100 kHz–10 mHz with a signal amplitude perturbation of 5 mV. The double-layer capacitance was calculated from the charge transfer resistance and characteristic frequency f_{max} in the semicircle maxima as:

$$C_{dl} = \frac{1}{2\pi f_{max} R_t} \tag{4.4}$$

The charge transfer resistance and efficiency increased, and the double-layer capacitance decreased with the higher inhibitor concentration. With a longer period of three hours of immersion in the presence of 80 mg/l inhibitor, the charge transfer resistance was nearly six times higher and the efficiency was as high as 90%. The decrease of the capacitance was attributed to two possible reasons of the decrease in the dielectric constant and/or the increase of the double layer thickness upon the adsorption of the inhibitor, in light of observations reported in [Popova, 2003, Oguzie, 2007]. The Nyquist plots increased in size with the inhibitor concentration with no indication that the interfacial interactions or the modes of the adsorption changed.

The authors found that EIS data fit the Temkin isotherm of the formula:

$$\exp\left(-2a\theta\right) = KC \tag{4.5}$$

θ conventionally represents surface coverage in relation to the inhibitor efficiency (η) as:

$$\theta = \frac{\eta\%}{100} \tag{4.6}$$

a is an interaction parameter, K is adsorption-desorption equilibrium constant, and C is the inhibitor concentration. The calculated values of (a) was −4.79 and K was 5.02×10^6 mol/l. The authors calculated the standard free energy of adsorption ($\Delta G^{\text{o}}_{\text{ads}}$) from the equation below, to be −48.18 kJ/mol:

$$K = \frac{1}{55.5}\exp\left(\frac{-\Delta G^{\text{o}}_{\text{ads}}}{RT}\right) \tag{4.7}$$

The authors ascribed the negative value of the interaction parameter to the interfacial repulsion as the inhibitor molecules adsorbed on the surface, and the relatively high value of K to the process of adsorption being more dominant than the desorption process, similarly to findings reported in [Refaey, 2004]. The negative value of $-\Delta G^{\text{o}}_{\text{ads}}$ indicated spontaneous adsorption.

Jiang *et al.* evaluated the inhibition performance of imidazoline introduced with different concentrations on N80 steel, in 3% NaCl CO_2-saturated solutions at 25 °C [Jiang, 2005A]. The tests were carried out at the OCPs of AC 5 mV voltage in a range of frequency from 1 mHz to 100 kHz. The authors noticed that the inhibitor produced two capacitive loops which increased in size with inhibitor concentration until beyond 100 mg/l (as an optimum concentration) without mentioning a reason. The inhibition data fit with Langmuir isotherm. The charge transfer resistance and inhibition efficiency increased with concentration to 100 mg/l, and the double-layer capacitance showed discrepancy.

Zhang *et al.* studied the inhibition performance of imidazoline-based inhibitor on CO_2 corrosion of a carbon steel in 3 wt% NaCl 60 °C solutions of two different microstructures [Zhang, 2018]. With the increase of concentration, the size of the Nyquist spectra increased in size, but with no indication from the Bode phase on a change in the inhibition mechanism or in the nature of the corrosion reactions, and the two microstructures exhibited the same EIS response. From the decrease in the double-layer capacitance, the authors linked it to the replacement of water molecules at the surface by the inhibitor molecules to adsorb. The equivalent circuit for all experimental conditions for both steels was simplistically proposed to be the same.

Lopez *et al.* evaluated three inhibitors, namely Aminopropylimidazole (API, analytical-reagent grade 98%; $R_1 = H$, $R_2 = (CH_2)_3{-}NH_2$) and two commercial imidazoline-based inhibitors (PC: $R_1 = (CH_2)_n{-}N+H_3$ $n = 8{-}12$, $R_2 = (CH_2)_2{-}NH_2$; and QB: $R_1 = (CH_2)_n{-}CH_3$ $n = 10{-}17$, $R_2 = (CH_2)_2{-}NH_2$) [Lopez, 2005]. The EIS response was measured at the corrosion potentials with an applied potential of $\pm 0.005\ V_{rms}$ and a frequency range of 20,000 to 0.05 Hz, at 2, 24, 48, 72, and 96 hours of exposure. The concentrations of API, PC, and QB were 100, 20, and 50 ppm, respectively. The authors postulated that there is no evidence for the formation of protective $FeCO_3$ for the appearance of one capacitive arc of a single time constant, so rather, there was a porous $FeCO_3$ layer of resistance that was much less than that of charge transfer resistance. For API inhibitor, the pyridine-like (sp^2) nitrogen atom would act as a Lewis base, while Fe^{3+}, Fe^{2+}, and metallic Fe would behave like acids acting as electron acceptors, with higher acidity corresponding to a higher oxidation state. The interaction at the surface was maintained by long-range electrostatic force, for the higher Lewis acidity of Fe_3C compared to metallic iron (α-Fe), providing preferential sites for the inhibitor to block the active sites. API did not have long hydrocarbon R_1 chains; therefore, EIS features corresponding to their blocking effect did not appear. For PC inhibitor, there was evidence of thicker $FeCO_3$ film formation, and it had R_1 chains, that was more stable than API as a barrier against the aggressive ions. The authors mentioned that the ammonium group in the chain could create inhomogeneous adsorbing layer due to the positive repulsive charge. As a result, Fe^{2+} could accumulate locally in a process governed by the microstructure. In the presence of QB, the corrosion attack was significantly less, with evidence from the Bode phase that the inhibition was two-time-constant based due to the long hydrocarbon chain R_1 with hydrophobic properties that allowed to adsorb as thermodynamically stable layer.

Olivares *et al.* monitored the changes at the interface of a mild steel immersed in CO_2-saturated 3% NaCl solutions with time with two consideration for two pH 3.8 and 5.5 values, in the presence of 30 ppm oleic imidazoline inhibitor [Zavala Olivares, 2007]. Represented in Nyquist, bode, and Bode phase plots, the effects at the interface were studied and modeled with an equivalent circuit. The behavior was two-time constant and the Nyquist curves increased in size. At frequencies higher than 100 Hz, the significance of time was more discerned. The authors added a distributed element in the equivalent circuit to account for the high-frequency response and the solution resistance did not show a change with time. The decrease in capacitance values at the interface indicated that the inhibitor protectiveness increased with time.

Belarbi *et al.* studied the effect of 100 ppm of 1-(2-aminoethyl)-2-oleyl-2-im idazolinium chloride on the corrosion of API 5L X65 carbon steel at 20 °C at OCPs. With the addition of the inhibitor, the capacitive loop was large without low-frequency inductive loop (attributed to the relaxation of the adsorbed chloride and hydrogen ions), and the Bode diagram indicated at a two-time constant response and inhibitor adsorption. The authors analyzed the capacitive loops evolving with time, finding a significance of the inhomogeneity of the surface, which seemed to increase in the presence of the inhibitor. The capacitance of the double layer decreased accordingly and the inhibitor as an adsorbed layer did not have a clearly measured dielectric property.

Singh *et al.* evaluated the interactions for J55 steel in the presence of three inhibitors AT-1, AT-2, and AT-3 at different concentrations [Singh, 2019]. The scans were at −250 mV to + 250 mV vs. OCP, at 1 mV/s, of 100 KHz to 10 mHz. At low concentrations, the spectra showed both capacitive and inductive characteristics, indicating the role of the inhibitor adsorption. Upon fitting the data, the charge transfer resistance and double-layer capacitance were of opposite trends in relation to the concentrations of all inhibitors. The maximum charge transfer resistance was by AT-1 due to OCH_3 functionality, with evidence of an adsorptive film of high homogeneity.

Zhang *et al.* examined the effects of thioureido imidazoline introduced in different concentrations on the interfacial interactions of X70 steel of 150-grit surface roughness [Zhang, 2019]. The influence of the inhibitor concentration was nearly independent of temperature and surface roughness. The polarization resistance increased and the constant phase element value decreased with the inhibitor concentration, indicating the inhibitor molecules upon adsorption on surface replaced water molecules in a single-time constant fashion.

Followed by stabilization of OCPs for nearly 60 minutes of immersion, Okafor *et al.* made EIS measurements of frequency range of 100 kHz–10 mHz with a signal amplitude perturbation of 5 mV, for N80 carbon steel in CO_2-saturated simulation formation solutions containing 50, 100, and 200 ppm RAIM [Okafor, 2011]. As shown in Figure 4.11, the spectra did not show much difference in shape, the high-frequency capacitive portions are due to the surface roughness and inhomogeneities, affecting the charge-transfer

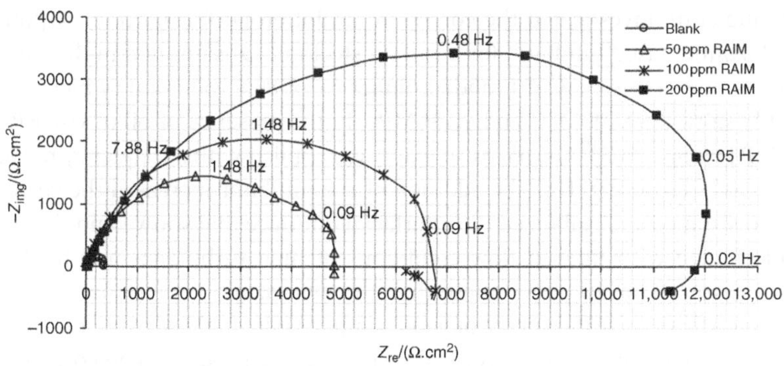

Figure 4.11 Nyquist plots for N80 carbon steel in CO_2-saturated simulated formation water in the absence (blank) and presence of RAIM at 20 °C. *Source:* Okafor [2011]/John Wiley & Sons.

resistance for dissolution, and the inductive loops were ascribed to adsorption of species that affect the negative potentials of the surface. The inhibitor, regardless of concentration, did not affect the mechanism of dissolution, but upon adsorption the charge-transfer resistance increased accordingly.

At OCPs, after 1.5 hours of immersion, Migahed *et al.* evaluated the inhibition performance of PQI and SQI in different concentrations from 50 to 500 ppm in oil well water formation solutions for X-65 carbon steel [Migahed, 2017]. The frequency range was from 100 kHz to 50 MHz using 10 steps per frequency decade at 25 °C. Charge transfer resistance, inhibitor film resistance, and inhibition efficiency increased, and the double-layer capacitance decreased with inhibitor concentration (which was attributed to replacement of water molecules with adsorbed inhibitor molecules). The authors modeled the inhibitor films as defective coating layers through which diffusion of ions are sustained after adsorption with time, indicated by diffusion tails in the Nyquist spectra. For calculating the double-layer capacitance and diffusion resistance, the authors utilized the following calculations:

$$C_{\mathrm{dl}} = \left(2\pi f_{\max} R_{\mathrm{ch}}\right)^{-1} \tag{4.8}$$

$$Z_{\mathrm{w}} = \sigma\,\omega^{-\frac{1}{2}}\left(1 - j\right) \tag{4.9}$$

$$\sigma = b\omega_o^{\frac{1}{2}} \tag{4.10}$$

where ($f_{\max}$) is the frequency at which impedance is maximum, (R_{ch}) is charge transfer resistance, (Z_{w}) is Warburg resistance, (ω_{o}) is Warburg coefficient, (σ) is the resistance of the diffusion through the inhibitor film.

Wang *et al.* investigated the corrosion inhibition performance of 50 ppm imidazoline-based inhibitor, for a carbon steel in CO_2-saturated multiphase 0.8 m/s flows (slug flow of Froude number 6), of 80% water cut (which was ASTM saltwater), at 40 °C and 0.136 MPa, in the presence of oil of viscosity of 2 cP [Wang, 2005]. A 4 mV AC voltage was applied with a frequency range from 100 KHz to 0.01 Hz. The addition of the inhibitor increased the corrosion resistance and stabilized the interfacial interface that showed no significant change with time in flow conditions of high turbulence and bubble distribution. The authors found that the polarization resistance values extracted from the EIS data are of good comparability to those extracted from electrochemical noise and electrical resistance experiments. The corrosion rates decreased from nearly 1 to 0.02 mm/yr upon the addition of the inhibitor.

Ortega-Toledo *et al.* evaluated the corrosion inhibition of commercial type HEI dissolved in pure 2-propanol in CO_2-saturated saltwater solutions at 50 °C, for X-80 pipeline steel, for concentrations of 5, 10, 25, 50, and 100 ppm [Ortega-Toledo, 2010]. The impedance spectra were produced after 30 minutes of immersion, at the OCPs, at a frequency range from 0.1 Hz to 30 kHz. For inhibitor concentrations less than 25 ppm, the spectra looked as single, depressed, capacitive arcs, with inductive loops at low frequencies associated with adsorbed intermediates from the anodic dissolution. At 25 ppm, the diffusion tale at intermediate and low frequencies indicated that the corrosion became more under diffusion control as the interactions sustain across the adsorbed inhibitor layer. At higher concentrations, the corrosion rate seemed to increase indicating that the 25 ppm is an optimum concentration.

Rivera-Grau *et al.* used 20 ppm organic HEI inhibitor to study the impedance of a 1018 carbon steel in 3% NaCl CO_2-saturated solutions, containing 10 vol.% diesel at 50 °C [Rivera-Grau, 2016]. The Nyquist arcs, which increased with time, were depressed and capacitive in high-frequency regions, followed by straight lines in the low-frequency regions, indicating diffusion-controlled processes.

For X-70 steel, Ortega-Sotelo *et al.* used commercial carboxyamido imidazoline, introduced in 1.6×10^{-5}, 3.32×10^{-5}, 8.1×10^{-5}, 1.6×10^{-4}, and 3.32×10^{-4} M, in CO_2-saturated 3% NaCl solutions at 50°C to evaluate the impedance response at OCPs [Ortega-Sotelo, 2011]. In 8.1×10^{-5} M solution, the mechanism of corrosion changed from being under mass transfer control to the mixed control of charge and mass transfer control to charge transfer control. At lower 3.32×10^{-5} inhibitor concentrations, the mechanism did not show change over time, and the size of the capacitive arcs was nearly three times smaller, indicating lower charge transfer resistance.

Porcayo-Calderon *et al.* tested 1018 carbon steel in heated 50 °C CO_2-saturated (3% NaCl + 10% diesel) emulsions, in the presence of 5, 10, 25, 50, and 100 ppm hydroxyethyl imidazoline derivate inhibitor, dissolved in pure 2-propanol [Porcayo-Calderon, 2015]. The measurements were carried out at OCPs with 10 mV sinusoidal excitation, across 0.01 Hz–100 KHz, with model PC4 300 Gamry potentiostat. The authors reported the evolution of thicker and increasingly protective inhibitor layer with time is manifested in the emergence of high-frequency bode phase peak within 24 hours, to be of maximum value of 75° phase angle at 10,000 Hz due to the oily imperfect nature of the adsorbed inhibitor layer. It should be noted that in inhibitor-free solutions, diesel did not induce such an effect (given its protective ability).

The oleic imidazolines formed a strong bonding by the head groups to the surface through the electron pair of nitrogen of imidazoline ring and self-assembly of the head groups to form a dense ordered layer. The inhibitor tails could have formed hydrophobic film impeding the migration of the corrosive species. The authors mentioned that the protectiveness of the inhibitor depended on its solubility, rate of transport toward the surface, and local pH. In the low-frequency regions, the charge-transfer, relaxation, and mass-transfer processes were postulated to take place within the pores of the adsorbed film and at the inhibitor-solution interface.

From Figure 4.12, the order of inhibitor efficiency was highest at 10, 50, 25, 100, to 5 ppm, attributed to the packing efficiency of the inhibitor molecules onto the surface. In low concentrations there could be unprotected sites, but in high concentrations the electrostatic repulsion amongst the adsorbed inhibitor molecules could prevent efficient inhibitor packing, causing unprotected sites to appear, thereby increasing dissolution rate.

4.1.3 Open-circuit Potentials

Open-circuit potential (OCP) monitoring is a simple yet effective technique for assessing the corrosion inhibition efficiency of imidazoline inhibitors. OCP measurements involve recording the potential of a metal electrode without imposing any external bias or current. By monitoring the changes in OCP over time, researchers can observe the inhibitor's ability to shift the OCP toward more noble values, indicating a reduction in the corrosion rate. OCP monitoring provides valuable information on the long-term performance, stability, and surface reactivity of the imidazoline inhibitor in oil flow environments.

Evaluating the efficiency of inhibition with electrochemical techniques requires a relative stability of the measured free potentials upon immersion of testing samples. Measuring the OCPs remains a convenient way to check

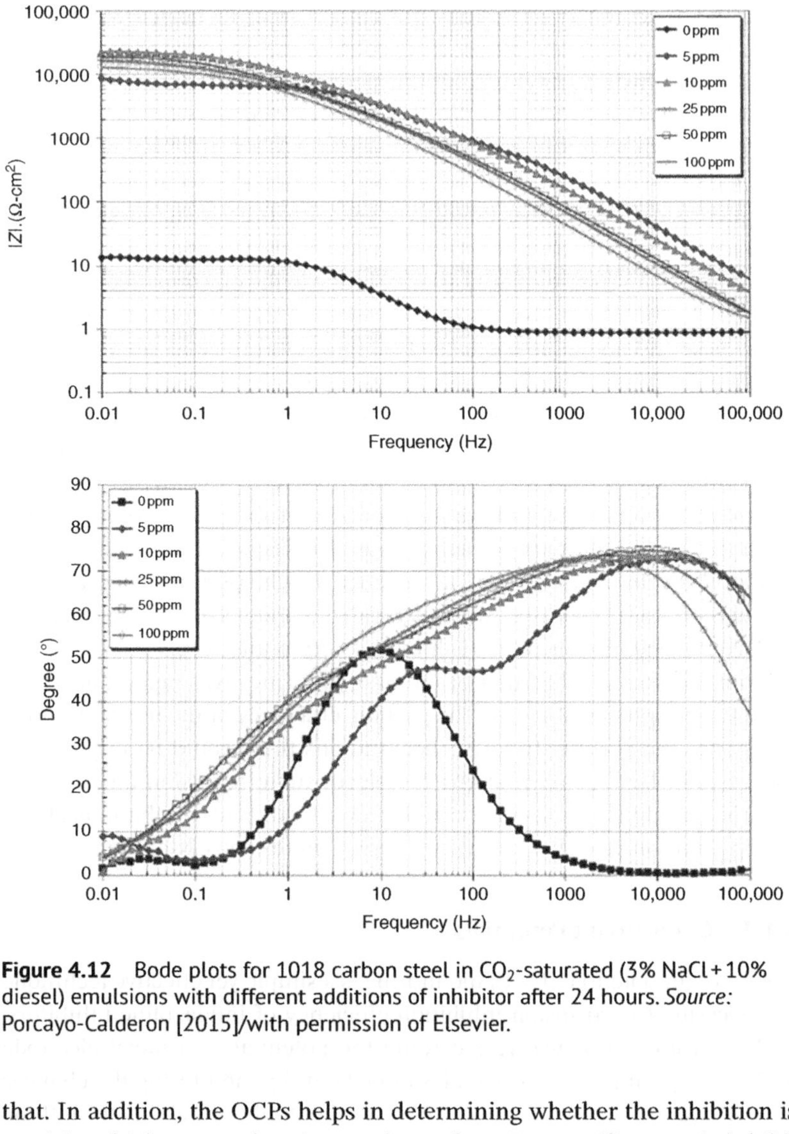

Figure 4.12 Bode plots for 1018 carbon steel in CO_2-saturated (3% NaCl + 10% diesel) emulsions with different additions of inhibitor after 24 hours. *Source:* Porcayo-Calderon [2015]/with permission of Elsevier.

that. In addition, the OCPs helps in determining whether the inhibition is partial or fairly comprehensive on the surface. For specific organic inhibitors, the OCPs helps in determining the threshold concentrations [Hassan, 2007]. The anodic inhibitors result in a positive shift of the measured free potentials. The variation of OCPs with time help in checking the stability of the measured potentials and degree of surface coverage before conducting electrochemical experiments that evaluated inhibition efficiency. The OCP measurements have sensitivity to environmental conditions and surface contamination. The long-term stability and reliability of OCP measurements can be affected by factors such as temperature fluctuations, dissolved oxygen concentration, and the presence of other corrosive species.

Desimone *et al.* evaluated the variation of corrosion potentials in saline CO_2-saturated solutions having different concentrations of AAOA; apparently reflecting the anodic characteristics that were proportional with the inhibitor concentration [Desimone, 2011]. In all concentrations, the OCPs became steady state, and all samples exhibited a "regular" attack, without specifying its nature.

In reference to Ag/AgCl/KCl reference electrode, Tantawy *et al.* studied the variation of OCPs during a relatively short 50 minutes' time, for three inhibitors in 3.5% NaCl CO_2-saturated solutions [Tantawy, 2020]. The three inhibitors, introduced in 1×10^{-3} M, were PGS-8, PGS-10, and PGS-12. The potentials exhibited a clear order of monitored OCPs within a short time, exhibiting distinct inhibitive capabilities without explanation, but the authors indicated the presence and role of the corrosion products at the interface with the adsorbed inhibitors. Li *et al.* studied the inhibition effect of 0.1 g/l 1-(2-aminoethyl)-2-heptadecyl imidazoline on the OCPs of N80 carbon steel in CO_2-saturated 1.65 wt% NaCl solutions containing acetic acid (HAc) of different concentrations [Li, 2017]. The investigation covered measuring the OCPs for connected surfaces inside and outside an artificial crevice of 12 mm × 12 mm with a thickness of 300 μm, for which the OCPs for solutions containing no inhibitor, and with inhibitor added in the beginning and after eight hours of the tests were monitored. In the presence of the inhibitor, the OCPs were noticeably higher, and they were higher outside the crevice than inside the crevice, linked to the gradual diffusion of the inhibitor into the crevice and the cathodic inhibition.

Zhang *et al.* examined the effects of surface roughness of 150 and 5000 grit of X70 steel in the presence of 0.488×10^{-5}, 1.22×10^{-5} and 1.22×10^{-4}, 2.44×10^{-4} mol/l of thioureido imidazoline for an immersion time of one hour [Zhang, 2019]. In the absence of the inhibitor, the OCPs did not show a difference, but with the addition of the inhibitor the OCPs were more positive and the smoother surfaces had OCPs more positive than the rough surfaces, indicating the better adsorbed film accordingly. The authors reported the same effects at 313 and 363 K with dependence on roughness and inhibitor concentration.

Belarbi *et al.* found for pre-corroded API 5L X65 steel, immersed for one hour in CO_2-saturated solutions containing 100 ppm of 1-(2-aminoethyl)-2-oleyl-2-imidazolinium chloride, a significant difference in the OCPs compared to those of the uninhibited solution [Belarbi, 2017]. In the presence of the inhibitor, the OCPs were as high as −0.62 V vs. Ag/AgCl, while in the absence of the inhibitor the OCPs were nearly −0.67 V vs. Ag/AgCl. In both solutions, the OCPs did not change much with time, and the OCPs of the inhibited solutions indicated that the inhibitor inhibited the anodic and cathodic reactions with effective adsorption on surface.

At 80 °C, Shamsa *et al.* examined the performance of optimum imidazoline concentrations on the corrosion performance of X65 steel in CO_2-saturated NaCl solutions, in relation to a short-term pre-corrosion period of two hours, during which the surface corrodes but without precipitation of corrosion products before injecting the inhibitor [Shamsa, 2020]. The OCPs increased from −690 mV to −650 and −625 mV, at 10 and 20 ppm, respectively, after 48 hours. At 30 ppm, the OCPs increased at two stages, first to −635 mV, stabilized for six hours before increasing again slowly to −600 mV, attributed by the authors to possible two-stage mechanism of adsorption and inhibitor/surface interaction. At concentrations greater than 40 ppm, the OCPs similarly indicated a more rapid film formation. The authors demonstrated the significance of pre-corrosion in the presence of 30 ppm inhibitor, showing that without pre-corrosion the OCPs stabilized within the first five hours of exposure, but with two hour pre-corrosion, the OCPs increased at two stages, taking longer 25 hours to stabilize, indicating that the pre-corrosion changes adsorption behavior for more effective film adsorption. The authors measured the changes in the pH values to not show significant change without pre-corrosion, while the inhibiting film formed more rapidly, than during pre-corrosion. The pH values showed greater increase in the experiment of pre-corrosion.

Khavasfar *et al.* monitored the variations in OCPs, in relation to concentrations of commercial-type imidazoline, for a mild carbon steel immersed in CO_2-saturated solutions containing 96.2 g/l NaCl, 3.3 g/l $CaCl_2$ and 1.8 g/l $MgCl_2$, proposed by NACE, as NACE1D-182 simulating gas well solution [Khavasfar, 2007]. In Figure 4.13, it shows that with increasing the concentration from 100 to 500 ppm, the OCPs increase, indicating that the inhibitor blocks the anodic active sites, for reactions dominated by the cathodic reactions.

Migahed *et al.* evaluated the inhibition performance of PQI and SQI in different concentrations from 50 to 500 ppm in oil well water formation solution, tabulated in the study for X-65 carbon steel [Migahed, 2017]. The authors analyzed the OCPs curves in the absence and presence of PQI inhibitor, which were similar in trend to those of SQI, for a time duration of 1.5 hours, at 25 °C. The authors attributed the decrease of potentials in the beginning to the breakdown of the air-formed oxide films, and with greater inhibitor concentration the OCPs were closer to the passivation potentials, to be of higher efficiency.

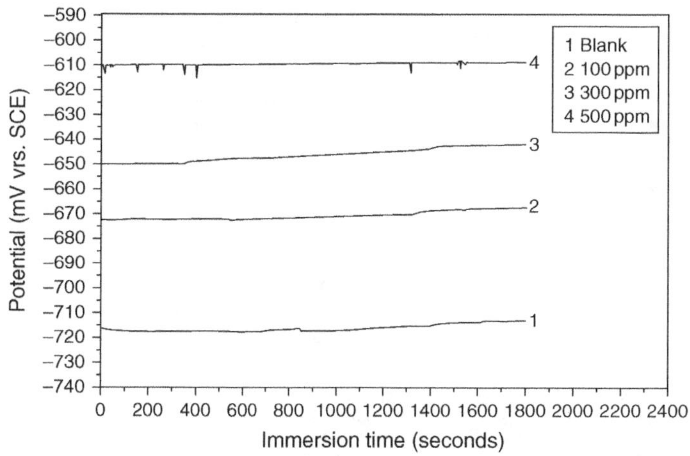

Figure 4.13 Variation of OCPs with immersion time containing different inhibitor concentrations at room temperature. *Source:* Khavasfar [2007]/MERC/CC BY 4.0.

4.1.4 Linear Polarization

Linear polarization is an electrochemical technique used to measure the polarization resistance and assess the inhibitory effects of imidazoline inhibitors on pipeline steel corrosion in oil flow environments. This technique involves applying a small potential perturbation around the corrosion potential and measuring the resulting current response. By analyzing the resulting polarization curve, researchers can determine the polarization resistance, which is inversely related to the corrosion rate. Linear polarization can provide insights into the inhibitive properties of imidazoline inhibitors, including their ability to reduce the corrosion current density, modify the corrosion kinetics, and mitigate the corrosion of pipeline steels. The accuracy of linear polarization results can also be influenced by factors such as electrode geometry, polarization time, and the presence of passive films on the metal surface.

In literature, linear polarization method covers both anodic and cathodic branches with a commonly selected rate of change of potential of 5 mV/s. Doležal *et al.* found that the corrosion rates, calculated from linear polarization and compared with data from EIS measurements, would not decrease after adding imidazoline with concentrations higher than 50 ppm [Doležal, 2013], as shown in Figure 4.14.

Ding *et al.* utilized linear polarization resistance to monitor CO_2 corrosion rates variation, of API 5L X65 carbon steel, with time in the absence and

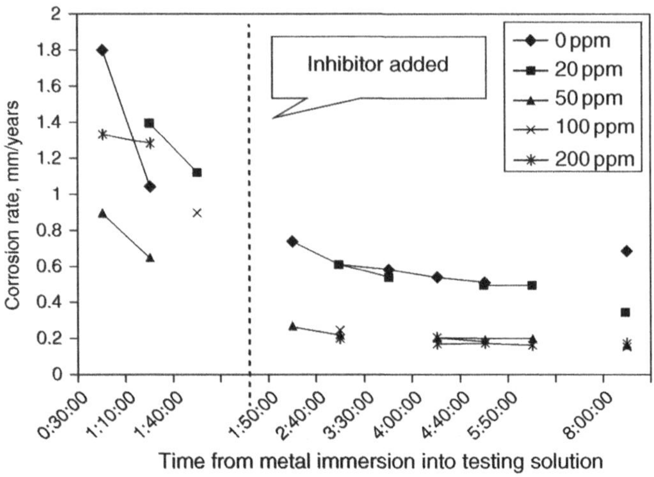

Figure 4.14 Corrosion rates calculated from linear polarization measurements for mild steel in CO_2-saturated brines having different concentrations of imidazolines. *Source:* Doležal [2013]/John Wiley & Sons.

presence of DETA/TOFA imidazoline, in a standard single autoclave system at 120 °C [Ding, 2018]. In the absence of an inhibitor the corrosion rates increased because of the high kinetics of dissolution and mass transfer at high temperature, before decreasing with time after 8 hours from nearly 10 to 1 mm/yr over 16 hours, because of the opposite kinetics of formation of protective corrosion products precipitating to limit dissolution and species transfer into and out of the interfacial layer, at which local concentration of Fe^{2+} ions and pH increase with time. With the injection of 440 ppm of the inhibitor, the corrosion rate quickly decreased to nearly 3 mm/yr, similar to previous results at lower temperatures, reported in [Ding, 2018].

In a two-autoclave system, the solutions and the specimens were heated separately in order to decrease the long heating and cooling transition periods during which the corrosion products could form before the injection of the inhibitor. The corrosion rate was initially 5 mm/yr, increased to and stabilized at 9 mm/yr. The behavior of the inhibited system was similar to that of the single-autoclave system at 120 °C, confirming the interfacial effects of the inhibitor, but the inhibition efficiency was calculated from the double-autoclave data, which were from a more reliable system. Belarbi *et al.* found for pre-corroded API 5L X65 steel, immersed for one hour in CO_2-saturated solutions containing 100 ppm of 1-(2-aminoethyl)-2-oleyl-2-imidazolinium chloride, a significant difference in the corrosion rates compared to those of the uninhibited solution [Belarbi, 2017]. In the presence of the inhibitor, the corrosion rates were as low as 0.03 mm/yr, while in the absence of the inhibitor the corrosion rates were 0.95 mm/yr.

Liu *et al.* evaluated the performance of two inhibitors AEI-11 and AQI-11 on CO_2 corrosion of N80 mild steel [Liu, 2009A]. The values of polarization resistance increased with time in the presence of the inhibitors, with indication that AQI-11 had better inhibition performance than AEI-11, confirming the result from other electrochemical tests.

Jovancicevic *et al.* in an early study evaluated the suitability of an imidazoline-based inhibitor with monitoring the corrosion rates and potentials variation with time in relation to the concentration [Jovancicevic, 1998]. As demonstrated in Figure 4.15, the corrosion rates were significantly less upon addition of the inhibitor, with dependence on concentration, and with anodic shifts that correlated to the formation of stable protective films. The

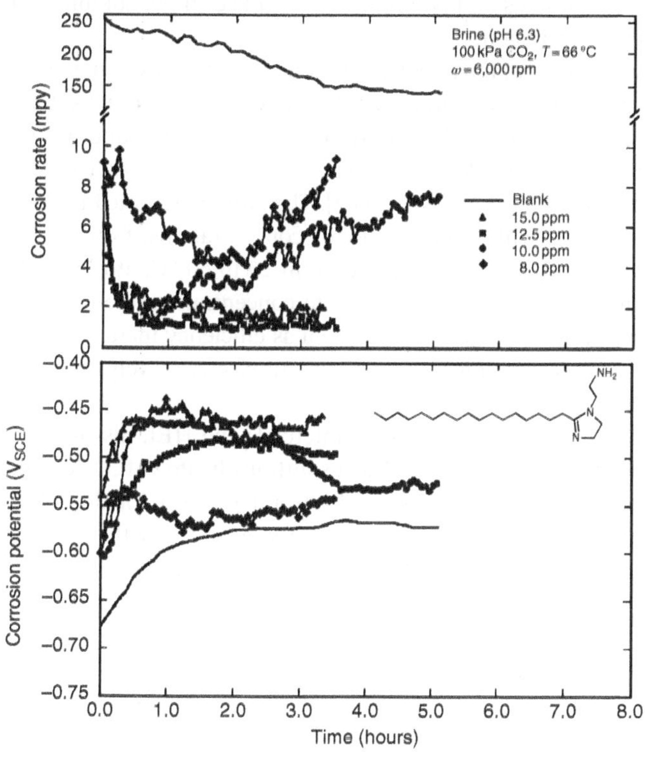

Figure 4.15 Corrosion rate/potential-time response for stearic imidazoline constant concentration treatment as a function of dosage. *Source:* Jovancicevic [1998].

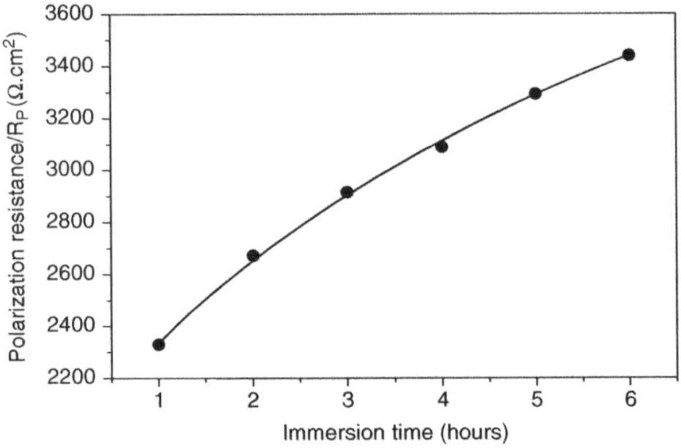

Figure 4.16 Variation of R_P value with immersion time in solution containing 300 ppm inhibitor at room temperature. *Source:* Khavasfar [2007]/MERC/CC BY 4.0.

variations in corrosion rates and potentials were linked to the solubility of the inhibitor and precorrosion.

Khavasfar *et al.* carried out linear polarization resistance measurements for a commercial-type imidazoline introduced in different concentrations, for a mild carbon steel immersed in CO_2-saturated solutions containing 96.2 g/l NaCl, 3.3 g/l $CaCl_2$ and 1.8 g/l $MgCl_2$, proposed by NACE, as NACE1D-182 simulating gas well solution [Khavasfar, 2007]. The authors reported the current densities decreased with higher inhibitor concentrations, from 100 to 500 ppm, and also with higher temperature, from 25 to 60 °C, attributed to enhanced inhibitor chemisorption. For 300 ppm concentration, the polarization resistance, as shown in Figure 4.16, as well as efficiency increased with immersion time, for polarization resistance measured successively after one hour during a six-hour immersion.

Ortega-Toledo *et al.* experimented a commercial type HEI, dissolved in pure 2-propanol, in CO_2-saturated 3% NaCl solutions heated at 50 °C. The inhibitor was introduced in 5, 10, 25, 50, and 100 ppm [Ortega-Toledo, 2010]. The polarization resistance increased (less corrosion rates) with inhibitor concentration and also with time. With time, the inhibition mode changed from active sites blocking effect (for short periods of time) to geometric blocking effect. With use of 20 ppm of organic HEI inhibitor, Rivera-Grau *et al.* monitored the change in polarization potential with time, for a 1018 carbon steel, in 3% NaCl CO_2-saturated solutions containing 10 vol.% diesel at 50 °C [Rivera-Grau, 2016]. The polarization resistance increased within 12 hours steadily to reach 50 kΩ, attributed to the inhibitor adsorption and formation of protective layer that was $FeCO_3$-based, precipitated from complex Fe^{2+} ions. R_p then decreased to 10 kΩ, without giving reasons.

Ortega-Sotelo *et al.* evaluated the effects of the commercial carboxy-amido imidazoline, introduced in 1.6×10^{-5}, 3.32×10^{-5}, 8.1×10^{-5}, 1.6×10^{-4}, and 3.32×10^{-4} M, in CO_2-saturated 3% NaCl solutions at 50 °C on the linear polarization of the X70 steel [Ortega-Sotelo, 2011]. The results indicated that the corrosion products were not protective, as the polarization resistance decreased with time in the inhibitor-free solution. The polarization resistance in all inhibitor-containing solution increased with time to reach steady values, indicating the effectiveness of the adsorbed inhibitor films.

4.1.5 Cyclic Voltammetry

Cyclic voltammetry (CV) is a versatile electrochemical technique used to evaluate the electrochemical behavior and performance of imidazoline inhibitors in oil flow environments. CV involves applying a potential waveform to the working electrode and measuring the resulting current response. By analyzing the shape, magnitude, and characteristics of the resulting voltammogram, researchers can obtain information about the redox behavior, adsorption/desorption processes, and stability of the imidazoline inhibitor. CV can provide insights into the mechanism of inhibition, the reversibility of electrode processes, the kinetics of the inhibitor-electrode interface, and the effectiveness of the imidazoline inhibitor under dynamic conditions. From voltammogram, there could be a lack of direct correlation between the observed peaks and the inhibitor's inhibitory performance. The choice of scan rate and potential range can also affect the results and should be carefully considered in experimental design.

Jevremović *et al.* investigated with CV the stability of TOFA/DETA imidazoline on a mild steel in CO_2-saturated 3 wt% NaCl brine, from −1 to −0.6 V vs. Ag/AgCl with a scan rate of 20 mV/s [Jevremović, 2016]. The criteria for the number of cycles was the stability in the voltammograms. Adding 90 ppm of the inhibitor decreased both the anodic and cathodic currents without showing special peaks or shoulders that would reflect specific reactions associated to the inhibitor, and that confirmed the physisorption nature of the inhibition mechanism. Feng *et al.* in saturated $Ca(OH)_2$ solutions of pH 12.5 studied the influence of a stearic imidazoline derivative, namely, 1-[*N,N*′-bis(hydroxylethylether)-aminoethyl]-2-stearicimidazoline (HASI) on the protection performance of an oxide film on a carbon steel [Feng, 2012]. As shown in Figure 4.17, adding 1000 mg/l inhibitor resulted in suppression of the anodic active and passivation currents as well as the cathodic currents. The oxidation peaks that correspond to the common reactions of the corrosion products of $Fe(OH)_2$ and $FeCO_3$ and the oxides were smaller. The authors

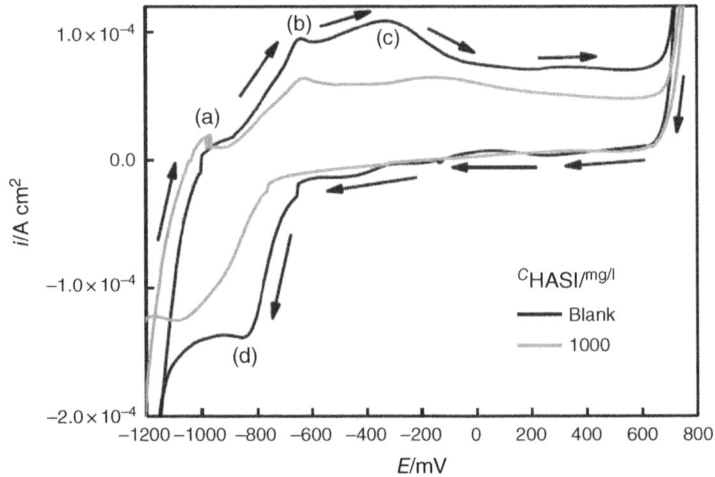

Figure 4.17 Cyclic voltammetry curves for carbon steel samples in saturated $Ca(OH)_2$ solutions without and with 1000 mg/l HASI. *Source:* Feng [2012]/with permission of Elsevier.

attributed that to the inhibitor hindering the formation of iron oxides and hydroxides as it adorbs onto the surface, and facilitating the ferrous species to oxidize to ferric ions.

4.1.6 Electrochemical Noise

Electrochemical noise analysis is a non-intrusive technique used to evaluate the corrosion behavior and performance of imidazoline inhibitors in oil flow environments. This technique involves measuring the spontaneous fluctuations in current or potential over time. By analyzing the noise signals using statistical and frequency-domain methods, researchers can extract valuable information about the corrosion rate, corrosion mechanism, and the effectiveness of the imidazoline inhibitor. Electrochemical noise analysis provides insights into the inhibitor's ability to suppress localized corrosion, modify the electrochemical reactions occurring at the metal surface, facilitate film formation, and overall corrosion inhibition performance in dynamic systems. It can also help identify and quantify various corrosion processes, such as pitting, crevice corrosion, and general corrosion, providing a comprehensive understanding of the inhibitory effects of imidazoline inhibitors.

Electrochemical noise technique evaluates the metastable pitting on corroding metals, with parameters including noise resistance (R_n), localization index (LI), and Hurst exponent (H). In relation to inhibitor concentration, and the environmental conditions, the parameters are used to elucidate the kinetics of dissolution, passive film formation, and inhibitor adsorption. In CO_2-saturated 0.2 mol/l NaCl + 0.2 mol/l $NaNO_2$ solutions, Zhao *et al.* used 50, 100, and 200 mg/l thioureido imidazoline (TAI) to inhibit the corrosion of N80 steel in electrochemical noise testing carried out after 24 hours of immersion. The DC trend, of the currents in Figure 4.18, was removed from the raw data by a least-square fitting method [Zhao, 2018]. The frequency of fluctuations was high, indicating the occurrence of metastable pitting. In addition, the susceptibility to localized corrosion was higher with the inhibitor concentration.

Ortega-Sotelo *et al.* reported, with the addition of 8.1×10^{-5} M of commercial carboxyamido imidazoline in a CO_2-saturated 3% NaCl solution at 50°C for X70 steel, low-intensity and high- frequency currents in the beginning of immersion, in Figure 4.19 [Ortega-Sotelo, 2011]. Compared to the uninhibited solution, the current intensity, although increased with higher rate of corrosion, was noticeably lower. The transient currents were of type indicative of the protective (yet not passive) role of a stable adsorbed layer.

R_n values obtained from the ratio between the potential noise standard deviation (σ_v) and the current noise standard deviation (σ_i), with time for different carboxyamido imidazoline concentrations, are shown in Figure 4.20 [Ortega-Sotelo, 2011]. They increased with time and reached semi steady values after four to six hours of immersion, in trends similar to the polarization resistance. The highest R_n was at 8.1×10^{-5} M, a concentration below and above which R_n was less.

Ortega-Toledo *et al.* examined the effect of flow with electrochemical noise on the inhibiting performance of 8.1×10^{-5} mol/l carboxyethyl-imidazoline in 3% NaCl CO_2-saturated solutions at 50 °C [Ortega-Toledo, 2010]. They evaluated the susceptibility toward localized corrosion, which decreased with rotation speed, in uninhibited and inhibited solutions in stationary and 2500 rpm conditions. That was indicated from transients changing from being of low intensity and high frequency to higher intensity and lower frequency. In the presence of the inhibitor, the transients decreased noticeably in frequency and intensity, but with rotation the inhibitor film showed evidence of breakdown from the transients of high frequency that also indicated localized corrosion [Hladky, 1982].

Wang *et al.* investigated the corrosion inhibition performance of 50 ppm imidazoline-based inhibitor, for a carbon steel in CO_2-saturated multiphase 0.8 m/s flows (slug flow of Froude number 6), of 80% water cut (which was ASTM saltwater), at 40 °C and 0.136 MPa, in the presence of oil of viscosity of 2 cP [Wang, 2005]. The authors studied the mean voltage and current readings and R_n and corrosion rate, with time for test solutions containing 50 and 100 ppm inhibitors, respectively. At 100 Hz sampling rate, the mean voltage

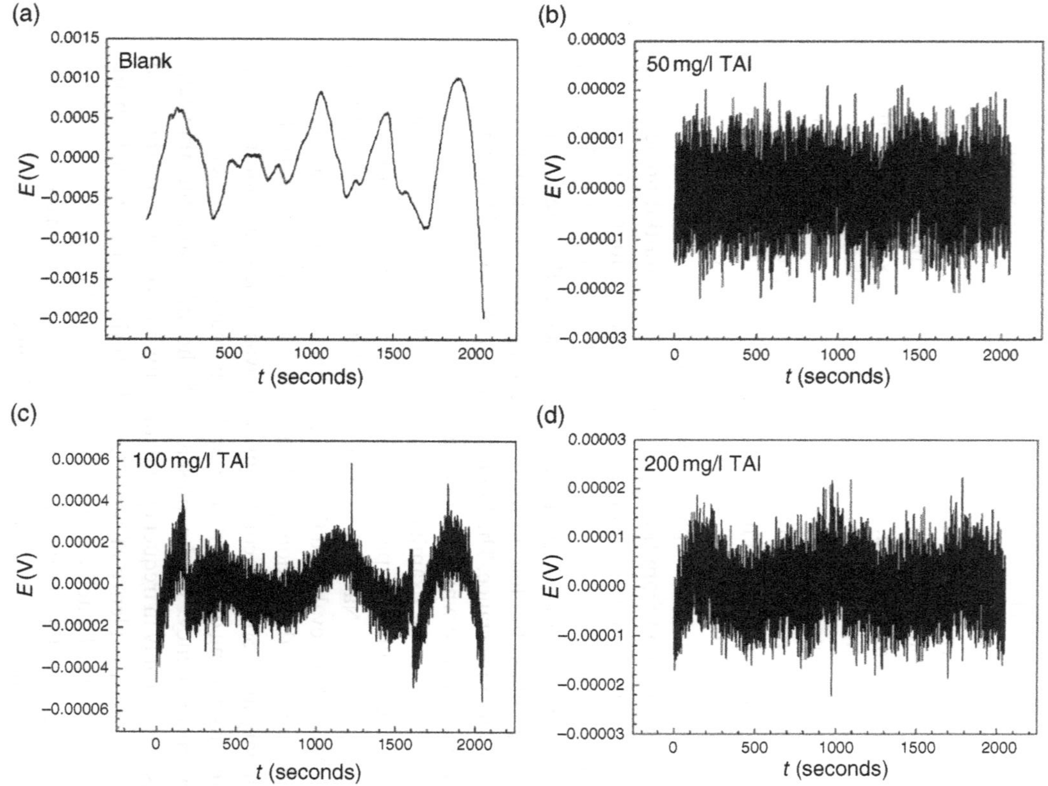

Figure 4.18 DC trend removed potentials vs time of N80 steel in the test solutions containing different concentrations of TAI. *Source:* Zhao [2018]/Elsevier/CC BY 4.0.

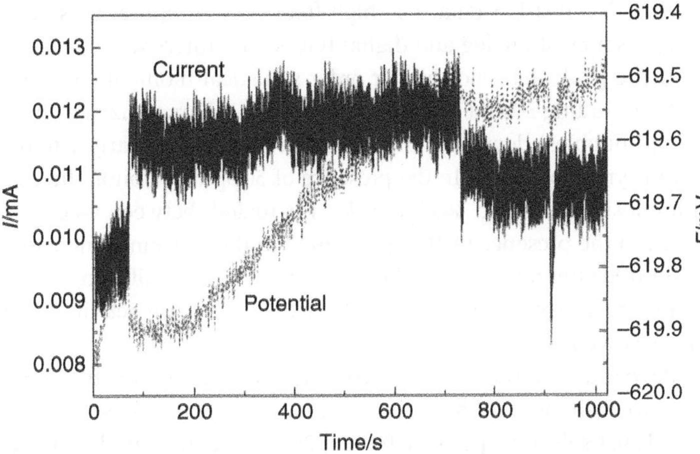

Figure 4.19 Noise in current and in potential for X-70 pipeline steel exposed to the CO_2-saturated 3% NaCl solution with the addition of 8.1×10^{-5} M of carboxyamido imidazoline. *Source:* Ortega-Sotelo [2011]/Springer Nature.

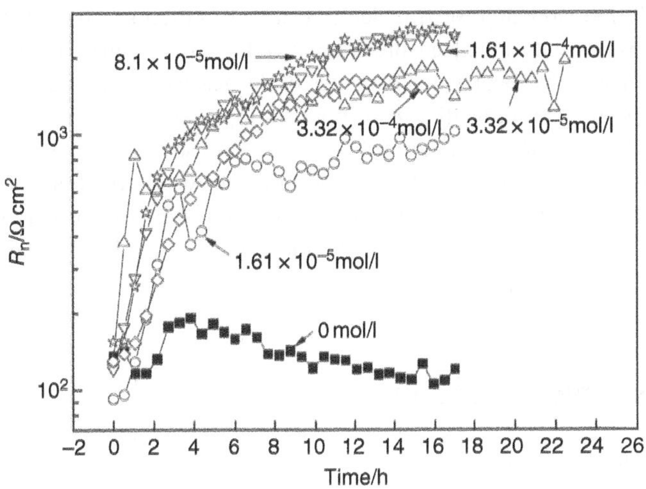

Figure 4.20 Effect of carboxyamido imidazoline concentration on the change in the R_n value with time for X-70 pipeline steel in CO_2-saturated 3% NaCl solution. *Source:* Ortega-Sotelo [2011]/Springer Nature.

and current were calculated using 50-seconds intervals, and in corrosion rate calculations, the cathode and anode Tafel slopes were 120 and 40 mV/decade, respectively. The noise data acquisition and analysis were carried out with a system of 1 μV and 100 fA resolution, and the filters were tuned to the sampling rate of Nyquist EIS data. The high-frequency computer noise was eliminated by a series of analog and digital filters. The filters were linear in phase, producing peak shapes of higher order statistical moments of minimum distortion. Adding 50 and 100 ppm to slug flows (characterized by high turbulence and bubbles) decreased the corrosion rates from nearly 1 to 0.5 and to 0.05 mm/yr, respectively. In the presence of 50 ppm inhibitor, the currents continuously decreased and the R_n fluctuated between 400 and 1200 ohm cm^2. In the presence of 100 ppm inhibitor, the resistance increased and then stabilized after nearly one hour, at 5000 ohm cm^2, manifesting a continuous inhibitor adsorption that decreased galvanic corrosion activities and with no rust observed.

Ortega-Toledo *et al.* used commercial type HEI, introduced in 5, 10, 25, 50, and 100 ppm, to evaluate it inhibiting corrosion of X-80 steel in 50 °C CO_2-saturated 3% NaCl solutions [Ortega-Toledo, 2010]. The potential and current fluctuations were recorded simultaneously, with THE use of automated zero resistance ammeter (ZRA) at a sampling rate of 1 point/s. In the beginning after introducing the inhibitor, high-frequency and high-intensity transients appeared, indicating the opposing kinetics of the corrosion and corrosion inhibition, and later the transients became less intense, attributed to pitting corrosion or to localized breakdown, followed by repair in the inhibitor film (driven by localized active sites underneath). With time, the spikes were less intense, indicating continuous suppression of corrosion. The authors calculated R_n by dividing the potential noise standard deviation, σ_v, over the current noise standard deviation, σ_i. R_n was comparable to the polarization resistance, obtained from polarization testing for corrosion rate calculation. The R_n increased with time and achieved steady-state values, and was highest in 25 ppm solutions.

4.2 Microscopic Evaluation

Microscopic techniques offer valuable insights into the microscopic evaluation of imidazoline inhibitors in oil flow environments. They allow researchers to visualize surface morphology, examine microstructural characteristics, assess local electrochemical behavior, and study the impact of inhibitors on corrosion processes. By integrating data from these techniques, researchers can gain a comprehensive understanding of the inhibitor's performance, surface interactions, and corrosion mitigation mechanisms. It is important to consider the limitations of each technique, such as the need for sample preparation, potential artifacts, interpretation challenges, and the inability to provide a complete picture of the chemical composition or dynamic behavior of the system.

4.2.1 Scanning Electron Microscopy (SEM) and Profilometry

Scanning electron microscopy (SEM), combined with profilometry in many recent studies, is a powerful technique used in the microscopic evaluation of imidazoline inhibitors' effectiveness in slowing corrosion of pipeline steels in oil flow environments. SEM utilizes a focused electron beam to scan the sample surface, generating high-resolution images. Profilometry, on the other hand, measures the surface topography and roughness. This combination allows researchers to visualize the morphology of the metal surface, including the presence of corrosion products, inhibitor films, and surface defects. Additionally, profilometry provides quantitative data on surface roughness, which can be used to assess the inhibitor's ability to modify the surface characteristics. These techniques provide valuable information on the surface morphology, inhibitor film formation, and the overall impact of the imidazoline inhibitor on the metal surface. SEM and profilometry are limited generally in providing information on the chemical composition and elemental mapping of the sample.

Jevremovic *et al.* evaluated the surface morphology of a mild steel surface after exposure of 24 h to CO_2-saturated 3 wt% NaCl solutions in the presence of 90 ppm of tall oil diethylenetriamine imidazoline (TOFA/DETA imidazoline), at 20 and 70 °C [Jevremović, 2016]. Adding the inhibitor inhibited significantly the corrosion reactions and the surfaces remained protected, compared to the surface in the absence of the inhibitor where they were covered with corrosion layers. Wang *et al.* studied the surface morphologies of J55 steel in high-pressure CO_2-saturated autoclave systems, at temperatures from 30 to 65 °C, in the presence of 80–150 ppm imidazoline inhibitor [Wang, 2014]. In the absence of the inhibitor, the corrosion rates decreased under the influence of the corrosion products proportionally with temperature higher than 60 °C, below which the corrosion products showed microcracks and were less protective – allowing diffusion of ferrous ions and anions, leading to high dissolution rates. In comparison with morphologies at same temperatures in uninhibited conditions, the amount of corrosion products were less in the presence of 80 ppm of the inhibitor, as seen in Figure 4.21, suggesting the effective influence of the inhibitor at different temperatures. The inhibitor adsorbed onto the surface as a monolayer that reduced the penetrability of the corrosive anodic and cathodic ions to the bare surface of the metal.

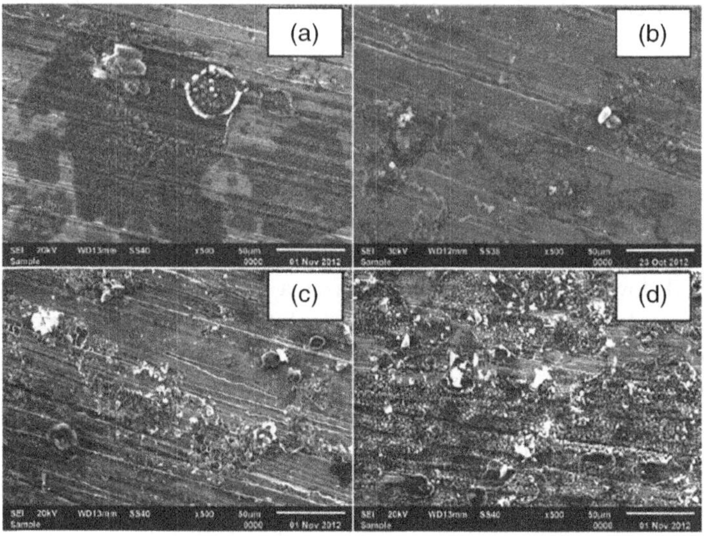

Figure 4.21 Surface morphologies of corrosion product films with effect of inhibitor (80 ppm) under simulated CO_2 flooding environment at (a) 30 °C; (b) 44 °C; (c) 55 °C; (d) 65 °C. *Source:* Wang [2014]/with permission from Taylor & Francis.

The authors correlated the corrosion efficiencies and nature of corrosion products to the changing solution chemistry of the produced water in the petroleum conditions simulated as test solutions. The water chemistry of changing concentrations of CO_2, bicarbonate, and carbonic acid was observed in relation to temperature and inhibitor concentration, as shown in Figure 4.22.

Tantawy *et al.* investigated the adsorption of novel synthesized cationic surfactant PGS-12 based on natural *Piper nigrum* on C1018 pipeline steel in 3.5% NaCl CO_2-saturated solution [Tantawy, 2020]. The surface morphology was studied with FE-SEM/EDX in the absence and presence of 1×10^{-3} M of PGS-12. In the absence of the inhibitor, the surface was heavily corroded and the spectrum showed the presence of Na, O, C, and Cl corresponding to $FeCl_2$, $FeCO_3$, and Fe_3O_4. In the presence of the inhibitor, the surface remained clean and smooth, and the peaks showed N and C, indicating the protective ability of PGS-12.

Ding *et al.* evaluated the morphology of the corrosion products of API 5L X65 carbon steel, in the absence and presence of DETA/TOFA imidazoline, of 440 ppm at 120 °C and 440 ppm and 880 ppm at 150 °C, in CO_2-saturated 1 wt% NaCl, in single- and double- autoclave system [Ding, 2018]. Both uninhibited and inhibited 120 °C samples showed extensive corrosion, but the features showed difference with crystalline and plate-like morphologies,

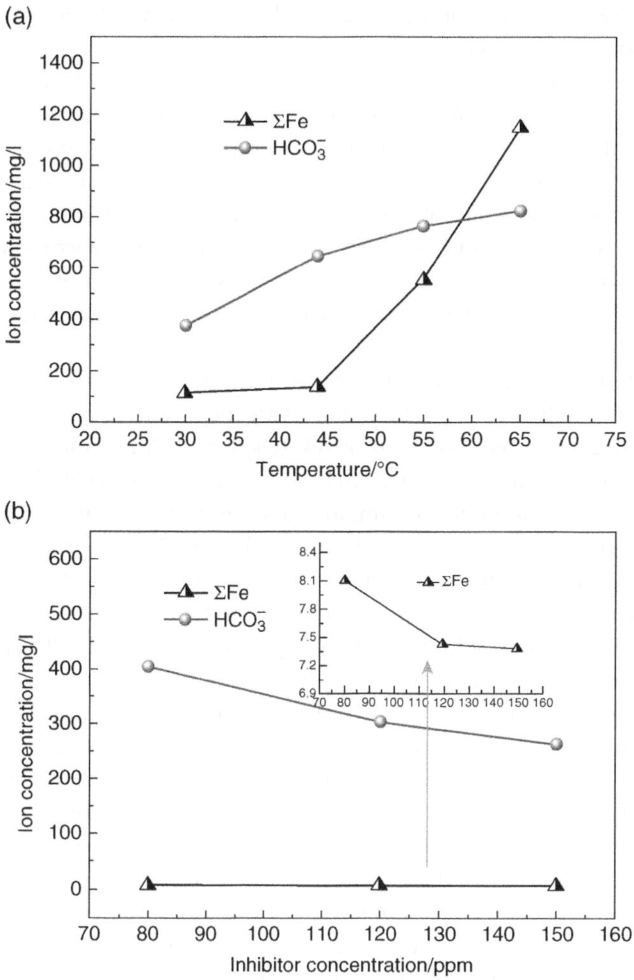

Figure 4.22 Variation characterization of measuring values of soluble iron ions and bicarbonate in relation to (a) temperature and (b) inhibitor concentration. *Source:* Wang [2014]/Taylor & Francis.

respectively. The corrosion products in the cross-sectional images showed that the inhibitor impeded the formation of a coherent, thick corrosion product layer. At 150 °C, the corrosion products were uniform and homogeneous; in the presence of the inhibitor, the corrosion products were porous and heterogeneous with no clear dependence on the inhibitor concentrations of 440 and 880 ppm. The cross sections showed that in uninhibited solution, the corrosion products grew to as much as 40 μm, but in the presence of the inhibitor at different concentrations, it seemed that it hindered the formation of thick, protective corrosion products.

In a double-autoclave system at 120 °C, the corrosion products were porous, indicating the delayed formation of the corrosion products. In the presence of the inhibitor, the morphology indicated the presence of iron carbide. At 150 °C, the corrosion products were uniform with crystalline structures in the absence of the inhibitor, but in the presence of the inhibitor, the corrosion products were less uniform. The cross sections of the corrosion products layers showed that the corrosion products were between 30 and 40 µm while in the presence of the inhibitor the corrosion layer thickness was very small.

Okafor *et al.* studied the inhibition performance of 2UEI for the corrosion of N80 mild steel. In the absence of the inhibitor, a very rough surface appeared after a seemingly general, not localized attack [Okafor, 2009]. Adding the inhibitor for three hours' immersion made the surface much less susceptible to corrode, indicating the effects of the inhibitive and protective ability of the inhibitor and the corrosion products in CO_2-saturated 3% NaCl solutions.

Li *et al.* investigated the inhibitive effects of 0.1 g/L1-(2-aminoethyl)-2-heptadecyl imidazoline on the corrosion products of N80 carbon steel in CO_2-saturated 1.65 wt% NaCl solutions containing 1 and 3 g/l HAc [Li, 2017]. The SEM was used to investigate the morphology of the corrosion products inside and outside an artificial crevice of 12 mm × 12 mm with a thickness of 300 µm. The authors studied the morphology of the products in solutions containing 1 g/l HAc without the inhibitor for 72 hours, and with the inhibitor added in the beginning and after 24 h of the experiments. Without the inhibitor, the corrosion was severe inside the crevice and the corrosion products outside the crevice were less than those inside the crevice, postulated to contain iron carbonate. Adding the inhibitor in the beginning limited noticeably the corrosion attack inside and outside the crevice, while adding the inhibitor after 24 hours was not effective; the corrosion products looked like those in the uninhibited solutions and they seemed to impede the adsorption of the inhibitor, especially inside the crevice. Interestingly, the corrosion products were less compact than those in the uninhibited solutions.

With the use of a 3D ultra-depth microscope, the corrosion surfaces were scanned to generate profiles and contours, shown in Figure 4.23, with respect to the surfaces studied by SEM after removal of the corrosion products. Without the inhibitor or with introducing the inhibitor after 24 hours caused the corrosion attack in the crevice to appear with grooves of appreciable depths, while introducing the inhibitor in the beginning preserved the surface of the crevice with nearly no variations in the depth. Without the inhibitor, the maximum depth was nearly 140 µm while, interestingly, the depth in the inhibited solution (inhibitor added after 24 hours) was 170 µm.

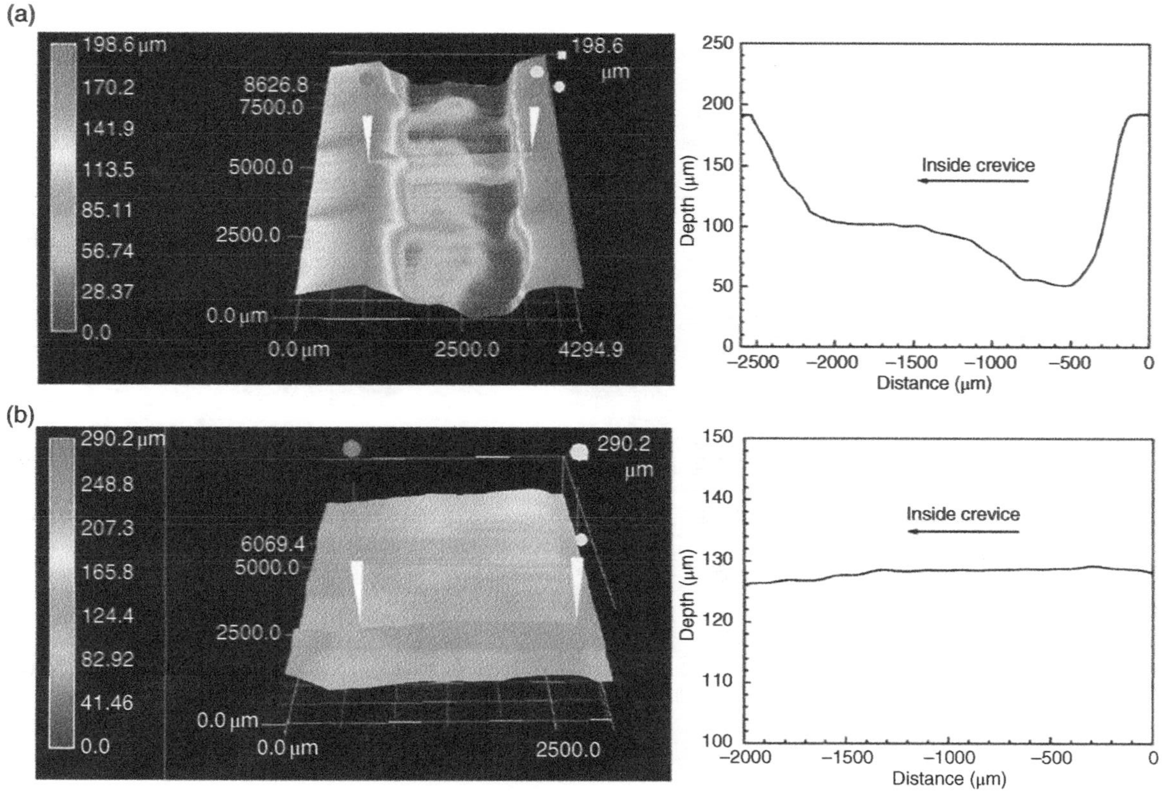

Figure 4.23 Profile at the crevice mouth of N80 carbon steel after corrosion in the CO_2-saturated 1.65 wt% NaCl solution containing 1 g/l HAc without or with IM for 72 hours: (a) without IM, (b) with IM added at the beginning, (c) with IM added after corrosion for 24 hours. *Source:* Li [2017]/with permission of Elsevier.

(c)

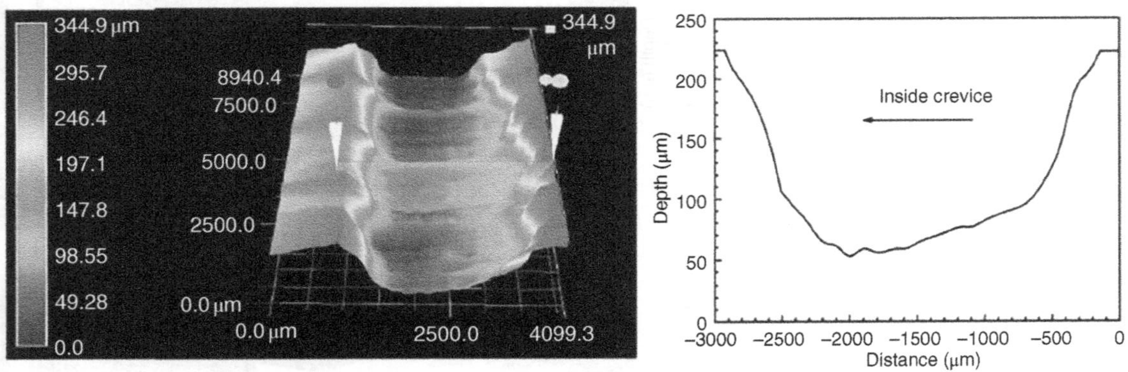

Figure 4.23 (Continued)

Zhang *et al.* examined with SEM the surface morphologies of a carbon steel with two different microstructures after corrosion of free immersion in 3 wt% NaCl, CO_2-saturated solutions at 60 °C [Zhang, 2018]. The first microstructure was produced by annealing, austenized at 850 °C and furnace cooled, having coarse laminar pearlite denoted as H steel. The second microstructure was produced by quenching and tempering, austenized at 850 °C, water quenched and tempered for one hour at 630 °C, having coarse laminar pearlite denoted as T steel. The authors considered two different immersion times, 24 and 240 hours, and two inhibitor concentrations 0.488×10^{-5} and 4.88×10^{-5} mol/l. In the absence of inhibitor, the morphologies after removal of the corrosion products showed that both microstructures were heavily, uniformly attacked.

With addition of inhibitor of 0.488×10^{-5} mol/l for 24 h, localized corrosion attacks took place, but with greater 4.88×10^{-5} mol/l inhibitor concentration, no localized attacks appeared. With longer immersion to 240 hours, both samples had localized attacks, more severe on H steel sample. A 3D laser microscope was used to measure the localized attack depths after scale removal, and the corrosion rate was calculated from an average of 10 pits of highest depths. With 0.488×10^{-5} mol/l and 24 hours of immersion, the localized corrosion rate of H steel was nearly 6 times greater than that of T steel. With immersion of 240 hours, the corrosion rates were higher and the attack mechanism was localized corrosion. SEM metallography for the samples of the pits in solutions of different concentrations of the inhibitor were presented in the study. H steel has more localized corrosion than T steel and the attack is mostly in the pearlite phase. The 3D surface morphologies of the localized attack in the presence of the inhibitor showed greater depth and diameter of the pits on H steel than on T steel.

Singh *et al.* evaluated the effects of three inhibitors AT-1, AT-2, and AT-3 on J55 in CO_2-saturated solutions [Singh, 2019]. Without inhibitor, the surface showed uneven corrosion morphology of multilayers. With the addition of 400 mg/l, AT-1 seemed to be most protective, exhibited from the smooth morphology.

Shamsa *et al.* evaluated the localized and general corrosion resistance of X65 carbon steel in high-temperature 80 °C, CO_2-saturated NaCl solutions, in the presence of different concentrations of imidazoline after pre-corrosion of 2 hours (short-term corrosion), for 48 hours of experimental time. The samples were studied with SEM and non-contact surface profilometry [Shamsa, 2020]. In the absence of inhibitor, the surface heavily corroded without precipitation of corrosion products. In the presence of 10 ppm inhibitor, the surface seemed to not be fully attacked, as it was partially protected from areas that showed roughness, with indication of vigorous dissolution partially. In the presence of 30 and 50 ppm inhibitors, the surface showed smoothness, consistent with the electrochemical results that the inhibitor at high concentrations provided enhanced protectiveness. The authors reported the significance of introducing the inhibitor at 30 ppm from the beginning of the experimental time without two-hour pre-corrosion. The surface showed that it was partially protected, as the inhibitor seemed to adsorb preferentially on the surface at relatively low concentrations. With pre-corrosion, the surface was more uniformly protected by the inhibitor.

The authors used an NPFLEX surface profilometer to study the depth and distribution of localized attacks during 48 h of immersion, including 2 hours of pre-corrosion, in the presence of 30 and 40 ppm of the inhibitor. The pit depths were nearly 25 μm at 30 ppm, while the maximum depths were 1.7 μm at 40 and 50 ppm. From the calculations of the localized inhibitor efficiency and generalized inhibitor efficiency, the authors reported optimum concentrations to control localized corrosion and generalized corrosion at 40 and 30 ppm, respectively. Distinct pitting attacks in the 30-ppm solution, but no pitting detected in the 40-ppm solution, were reported.

Xinag *et al.* studied the corrosion inhibition of N80 steel in supercritical CO_2 conditions, containing SO_2, NO_2, and O_2 impurities in the presence of 300 ppm imidazoline, at 10 MPa and 60 °C, immersed for 24 hours [Xiang, 2017]. The surfaces were covered with adherent corrosion products but had cracks, indicating low protectiveness. Using profilometry for samples of removed corrosion products, they reported pitting corrosion attacks, attributed to non-uniform adsorption of the inhibitor molecules on the surface. The authors calculated the corrosion rate to nearly be 6.20 mm/yr. From high-resolution SEM morphologies for EDS analysis, the authors reported the presence of Fe_3C. Migahed *et al.* evaluated the inhibition performance of SQI, introduced in 500 ppm to oil well water formation solution, in which a sample of X-65 carbon steel was immersed for 60 days [Migahed, 2017]. The authors conducted SEM and EDX analyses for samples immersed in inhibitor-free and inhibitor-containing solutions. In the inhibitor-free solution, the surface was vigorously attacked, but in the presence of the inhibitor the surface was protected with no appearance of pit or localized attacks. The EDX did not indicate significant influence of the inhibitor on the detected constituents of the corrosion films.

Porcayo-Calderon *et al.* examined the surface morphology of carbon steel immersed in CO_2-saturated emulsions containing 3% NaCl and 10% diesel, in the presence of hydroxyethyl-imidazoline derivate inhibitor in different concentrations, shown in Figure 4.24 [Porcayo-Calderon, 2015]. In the absence of the inhibitor, the surface was rough and had localized corrosion pits. With the addition of 5 ppm inhibitor the surface had pitting, and by increasing the concentration to 10 ppm, general corrosion occurred without pitting. With the increase of concentration to 25, 50, and 100 ppm, pitting appeared.

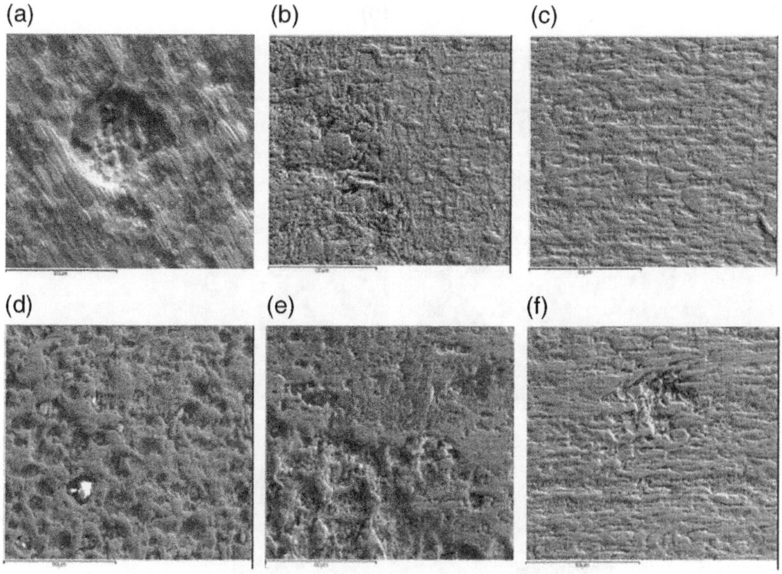

Figure 4.24 Corrosion morphology of carbon steel in CO_2-saturated (3% NaCl + 10% diesel) emulsions with the inhibitor in concentrations of (a) 0 ppm, (b) 5 ppm, (c) 10 ppm, (d) 25 ppm, (e) 50 ppm, and (f) 100 ppm. *Source:* Porcayo-Calderon et al., [2015]/with permission of Elsevier.

The authors attributed the behavior to high concentration of adsorbed inhibitor molecules having electrostatic repulsion, causing surface areas becoming unprotected and therefore prone to localized attack.

4.2.2 Transmission Electron Microscopy (TEM)

Transmission electron microscopy (TEM) is a high-resolution imaging technique used to study the microstructure and elemental composition of materials at the nanoscale. In the evaluation of imidazoline inhibitors, TEM provides detailed information about the crystallographic structure, grain boundaries, and the presence of corrosion products at the atomic level. By transmitting an electron beam through a thin sample, TEM generates bright-field and dark-field images, selected area electron diffraction patterns, and energy-dispersive X-ray spectroscopy data. These data help identify the inhibitor's impact on the microstructure, crystal defects, and the formation of protective layers. TEM requires sample preparation techniques that can introduce artifacts and may not be suitable for in situ or real-time observations.

Zhang *et al.* used TEM to study the dislocations and nucleation of cracks in relation to the microstructure of two heat-treated carbon steel samples in

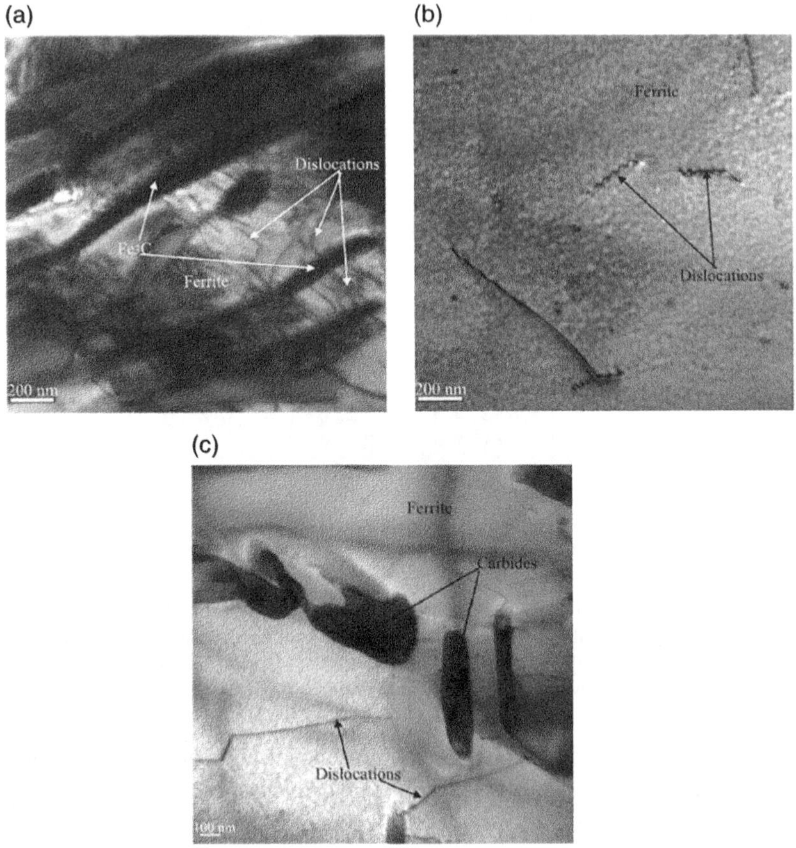

Figure 4.25 TEM images of the studied steels: (a) dislocations in pearlite region in H steel; (b) dislocations in ferrite region in H steel; (c) dislocations in T steel. *Source:* Zhang et al., [2018]/with permission from Elsevier.

CO_2-saturated solutions, in the presence of imidazoline-based inhibitor [Zhang, 2018]. The dislocations were linked to the pearlite phase, shown in Figure 4.25, which contributed to the variation in the corrosion rates variably in the absence and presence of the inhibitor, in different concentrations, with different immersion times. The authors, however, did not study the dislocations for samples immersed in inhibited solutions.

4.2.3 Scanning Kelvin Probe Force Microscopy

The principle of SKPFM is associated with an AFM tip with an applied bias, scanning a sample surface to generate a map of different potentials in reference to morphology/roughness of a surface. The relationship between the detected potential difference (V) and electron work function (EWF), which is defined as the minimum energy required to remove an electron from a surface (Fermi level) to a point just outside the metal, is as follows [Li, 2006, Ravi, 2018]:

$$V = \frac{\varnothing_{tip} - \varnothing_{sample}}{e^-} \tag{4.11}$$

where $\varnothing_{tip}$ is the EWF of the tip, which is 5.5 eV for Pt/Ir, and $\varnothing_{sample}$ is the EWF of the sample.

Zhang *et al.* found that in the presence of 1.22×10^{-5} mol/l and 2.44×10^{-4} mol/l of thioureido imidazoline in CO_2-saturated solutions, the Volta potential differences of 150-grit surfaces are higher than 5000-grit surfaces after immersion for 5 minutes [Zhang, 2019]. The authors reported that higher surface roughness decreased EWF, and the Volta potential difference was inversely proportional to EWF, attributed to an increase in the corrosion rate.

In the presence of low concentration of the inhibitor, the molecules would adsorb better on the smoother surfaces (of higher EWF), and would cover completely the rough surfaces. In the presence of high concentration of the inhibitor, the molecules would cover both the smooth and rough surfaces, but the rough surfaces (of lower EWF) were susceptible to localized corrosion.

Zhang *et al.* used SKPFM with an atomic force microscope (MFP-3D Infinity) and SCM-PIT plated Pt/Ir nanoprobe to measure the surface potential of steel samples studied for corrosion and inhibition performance in CO_2-saturated solutions at 60 °C [Zhang, 2018]. Topography and Volta potential were sampled with 256 × 256 pixel density and scan frequency of 1 Hz. The specimens were successively abraded with 150-, 400-, 800-, 1500-, 2000-, and 5000-grit SiC papers, then polished with 2.5-μm diamond abrasive, cleaned with distilled water and alcohol. The samples were electro-polished (solution, glycerol : perchloric acid : alcohol = 0.5 : 1 : 8.5 at room temperature, electropolish voltage: 30 V, electropolish time: 35 seconds), dried with alcohol. SKPFM tests were carried out at room temperature in open air. Figure 4.26 shows the initial topography and Volta potential of the samples before immersion. The images show lamellar pearlite and ferrite (bright and higher Volta potentials). The Volta potential difference between Fe_3C and ferrite in the pearlite areas was smaller than that between pearlite and ferrite, and the Volta potential differences between the granular pearlite and the ferrite was nearly 5 mV. The surfaces were measured after immersion for 10 minutes in CO_2-saturated 3 wt% NaCl solution with 0.488×10^{-5} mol/l inhibitor at room temperature. NaCl particles were observed on the surface and the Volta potentials after immersion were higher than before immersion, attributed to the adsorption of the inhibitor. Ferrite had higher Volta potentials than pearlite and Fe_3C had higher potentials than both, and Volta potential differences between granular pearlite and ferrite increased after immersion. SKPFM

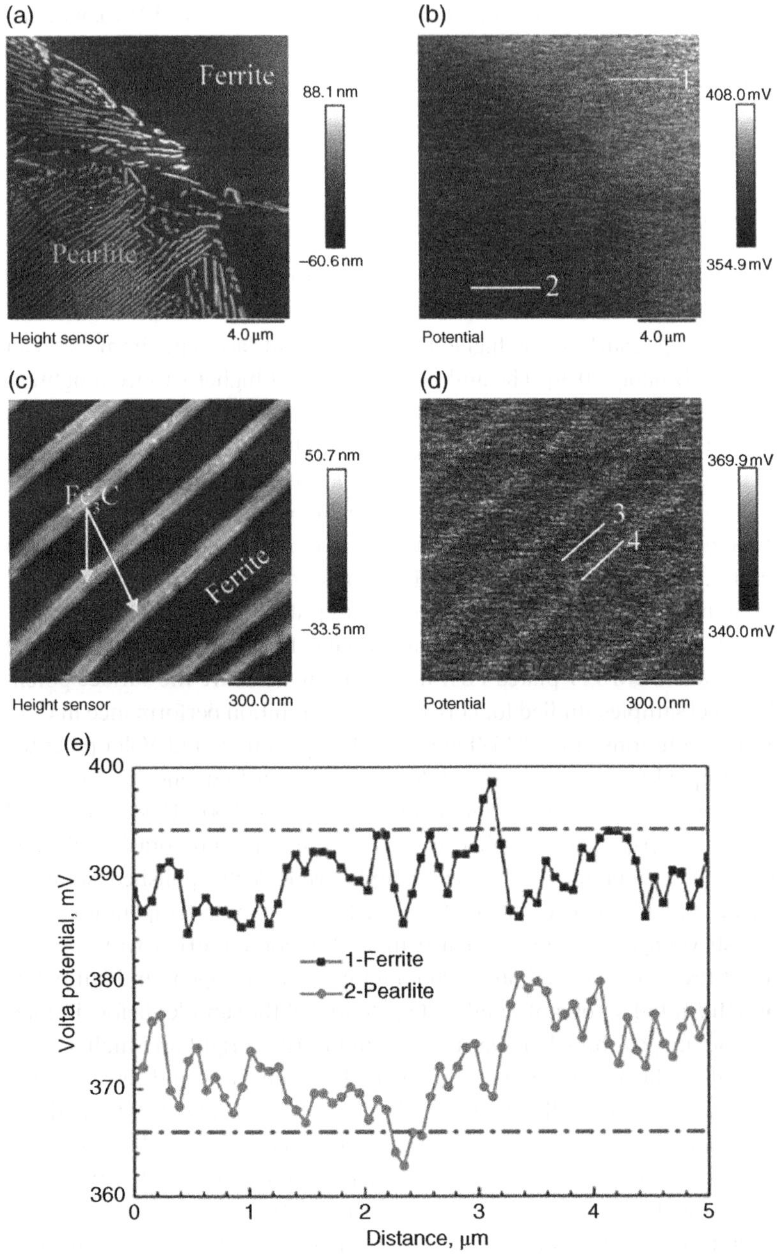

Figure 4.26 Surface topography and Volta potential mapping images of electropolished H steel sample before immersion measured by SKPFM: (a, b) lamellar pearlite and ferrite; (c, d) lamellar pearlite; (e) the line analysis of surface potential in (b); (f) the line analysis of surface potential in (d). *Source:* Zhang et al., [2018]/ with permission from Elsevier.

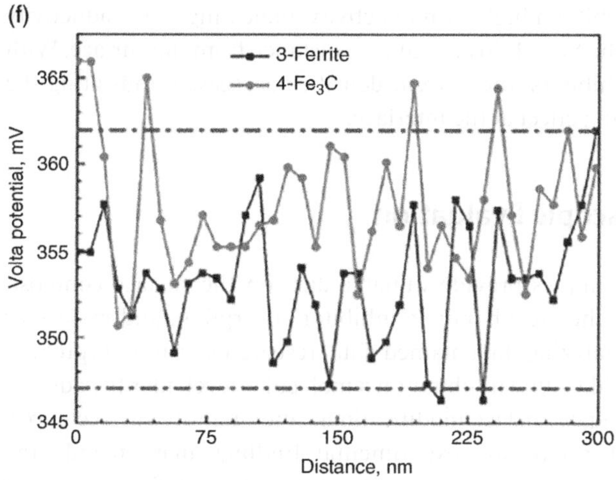

Figure 4.26 (Continued)

showed evidence for uniform inhibitor adsorption with strong hydrophobic capacity, to an extent the size of deposit NaCl particles at nanoscale changed.

4.2.4 Scanning Electrochemical Microscopy

Scanning Electrochemical Microscopy (SECM) is a powerful technique for assessing the effectiveness of corrosion inhibitors in steel at a microscopic level. The experimental setup involves a three-electrode system, featuring a working electrode, a reference electrode, and a counter electrode, along with a scanning system and a feedback mechanism. The SECM probe, often a microelectrode, is positioned near the steel specimen, and the instrument scans it in two dimensions to create electrochemical maps, revealing spatial variations in corrosion activity. The scan rate, tip-substrate distance, and applied potential or current are critical parameters influencing the resolution and reliability of the obtained data. The SECM analysis includes the interpretation of corrosion maps, which illustrate variations in corrosion rates across the sample. Additionally, researchers examine current-voltage curves at specific points to analyze the electrochemical behavior of the steel in the presence of inhibitors. The feedback control system is crucial for maintaining a constant tip-substrate distance during the scanning process. SECM provides high spatial resolution and real-time monitoring capabilities, making it useful tool for understanding the local corrosion behavior of steel surfaces and evaluating the performance of corrosion inhibitors.

Singh *et al.* examined the inhibition performance of three inhibitors AT-1, AT-2, and AT-3 on J55 steel [Singh, 2019]. The authors found first without addition of an inhibitor a high current activity, indicating the conductivity of the surface for the SECM tip at a suitable distance from the surface. With the addition of inhibitors, the current densities decreased, indicating the inhibitors' insulating effect at the interface.

4.3 Spectroscopic Evaluation

Spectroscopic techniques provide valuable data on the surface composition, morphology, chemical bonding, inhibitor adsorption, and molecular interactions. By analyzing the obtained data, researchers can interpret the behavior of the imidazoline inhibitor on a molecular level, identify adsorption modes, investigate surface modifications, and quantify surface coverage. Examples of interesting experimental findings may include the determination of inhibitor adsorption strengths, visualization of surface morphological changes, and identification of specific molecular interactions. However, limitations exist, such as the surface sensitivity of certain techniques, limitations in depth profiling capabilities, potential artifacts from tip-sample interactions, and challenges in quantifying coverage and thin film analysis.

4.3.1 X-ray Photoelectron Spectroscopy

X-ray photoelectron spectroscopy (XPS) is a powerful experimental technique used to analyze the surface composition, oxidation states, and chemical bonding of imidazoline inhibitors in oil flow environments. By irradiating the sample with X-rays, XPS detects and measures the emitted photoelectrons, providing valuable information about the inhibitor's molecular behavior. This technique can identify the types and concentrations of elements present on the surface, allowing for the characterization of inhibitor adsorption and the formation of protective layers. XPS data can be used to interpret the inhibition mechanism, assess the inhibitor's surface coverage, and monitor changes in the surface chemistry over time. However, XPS only provides surface-sensitive information and may not capture the full depth profile of the sample.

The nitrogen component of imidazolines is generally difficult to detect by XPS for corrosion products immersed for a long time in CO_2-saturated solutions [Zhang, 2001]. Rather, the role of some of the organic compounds in the inhibitor was postulated to affect the roles of H_2O and O_2 molecules, affecting ultimately the oxide constituents of the corrosion films and their corresponding peaks. Singh *et al.* examined the inhibitive effects of chromeno pyrimidine (CP) derivatives on the corrosion of N80 steel in solutions containing

15% HCl [Singh, 2020]. The XPS spectra were resolved into the spectra of components shown in Figure 4.27. Based on previous findings [Bentiss, 2009], the first peak at 284.38 eV indicated C—C, C=C and C—H bonds of the aromatic ring, and the second peak at 284.38 eV indicated C—N, C=N and C—O bonds of the pyrimidine ring [Arukalam, 2015]. The N1s spectrum confirmed a CP-1 film formation on the metal surface of a direct bond formation between the CP-1 nitrogen and iron atom of the N80 steel, as the first peak at 398.52 eV indicated at the C—N bond and the second peak at 399.47 eV indicated the non-protonated N atoms (¼N-structure) in the pyrimidine ring [Kang, 1992], and the third peak at 400.49 eV indicated adsorption of nitrogen atom of the pyrimidine ring with the surface. For O1s spectra, the peaks at 530.67 and 531.56 eV indicated the bonding of oxygen and OH^-, respectively with the

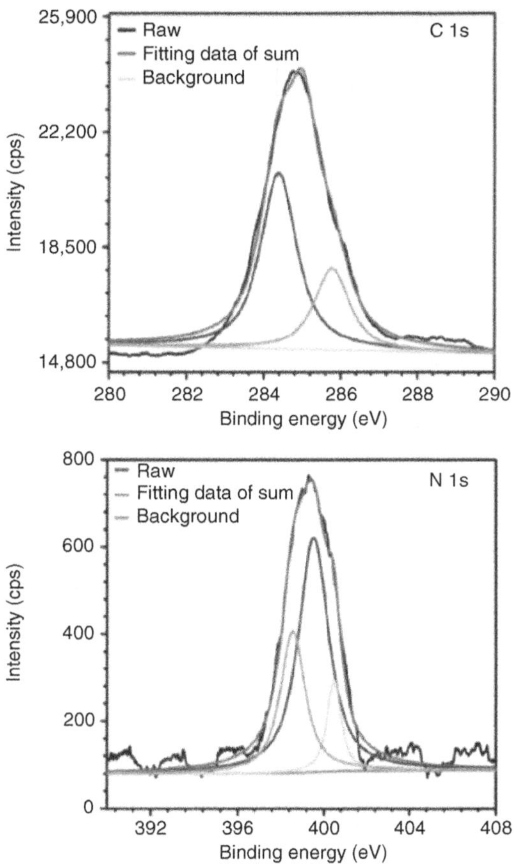

Figure 4.27 XPS spectra images of adsorbed CP-1 molecule on N80 steel. *Source:* Singh [2020]/with permission from Elsevier.

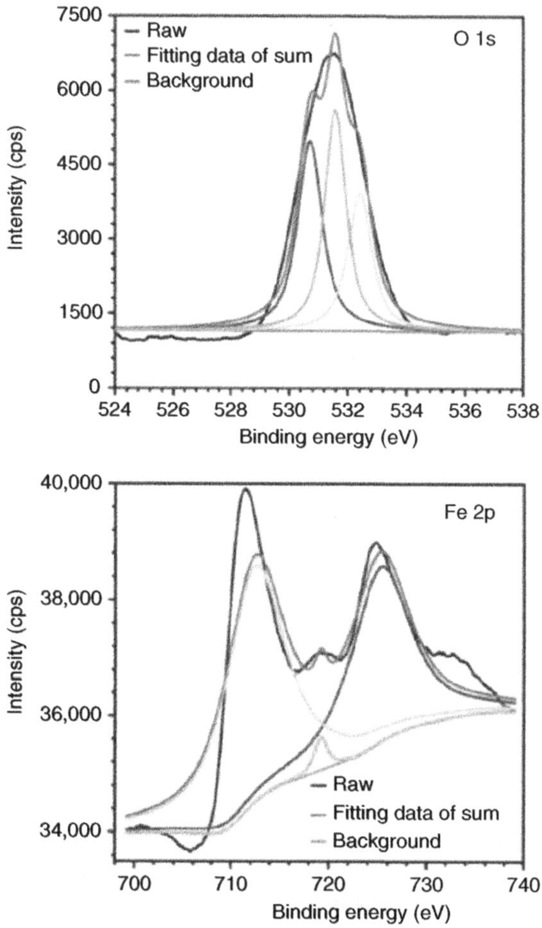

Figure 4.27 (Continued)

surface as Fe_2O_3 and/or Fe_3O_4 oxides, and hydrous oxides of the iron oxides such as FeOOH [Temesghen, 2002], and the third peak at 532.41 eV indicated at adsorbed water and oxygen [Babić-Samardžija, 2005]. Fe2p spectrum indicated the oxidation of the surface to different products.

Wang *et al.* utilized XPS to study Q235 steel covered with TAI, introduced in different concentrations, after 10 hours of immersion in CO_2-saturated 2% NaCl solution [Wang, 2011]. Oxygen and carbon showed evidence of the highest amounts and nitrogen and iron (inevitable to oxidize) of small peak intensities. Carbon and nitrogen were attributed by the authors to the TAI inhibitor, sulfur was not detected, and the binding energies (corrected for charging effects) showed slight dependence on the inhibitor concentration, making 0.15 mmol/dm^3 a representative sample for the high-resolution spectra. From the binding energy (nearly 400 eV, and higher than that of pure imidazoline [Jing-mao, 2004]) of N1s at different concentrations of the inhibitor, there was an established evidence for adsorption. While the binding energies of Fe 2p3/2 at different inhibitor concentrations were lower than that of the uninhibited sample, the authors concluded that the inhibitor was chemically adsorbed. At 0.31 mmol/dm^3, no nitrogen was detected, attributed to the perpendicular adsorption of the hydrocarbon chain of TAI, at lengths of higher than 10 Å, which is the depth inspection limit of the XPS instrument used in the experiments, similar to a previous finding [Zhang, 2001].

Zhao *et al.* studied the combined effects of oleic-based imidazoline (OIM) and sodium benzoate (SB) on inhibiting corrosion of a mild steel in CO_2-saturated solutions [Zhao, 2012]. They studied in detail the significance of inhibition by XPS, in the presence of 10 mg/l OIM + 10 mg/l SB, at10 and 30 min of immersion times. Oxygen and carbon showed the highest intensities and nitrogen and iron appeared with small intensities. The peaks of carbon and nitrogen are linked to the OIM. At higher resolution, N1s spectrum with a binding energy of 399.6 eV indicated the carbon-nitrogen bond in the imidazoline molecule, and the area of the spectrum increased with time, indicating more inhibitor molecules to adsorb onto the surface.

The authors highlighted the Fe2p3/2 line deconvolution for immersion times of 10 and 30 minutes at higher resolution, with indication of two chemical states FeOOH (711.8 eV) and $FeCO_3$ (710.2 eV), cited from [Heuer, 1999] – with notation that they result from the exposure from the original $FeCO_3$ product to air (hydrolysis and oxidation). The immersion time did not seem to induce significant difference. Oxygen, from O1s line deconvolution, appeared in three chemical states for 10 minutes of immersion, FeO*OH (530.1 eV), FeOO*H (531.2 eV) and $FeCO_3$ (531.9 eV), but for 30 minutes of immersion, it appeared in five chemical states, $FeCO_3$ (531.9 eV), FeO*OH (530.1 eV), FeOO*H (531.2 eV), phCOONa (531.4 eV), and phCOO*H (533.4 eV), which exists in less amount than phCOONa. From the XPS observations, the authors indicated that it could be sodium benzoate or benzoic acid molecules that existed in the adsorbed film at longer 30 minutes, not 10 minutes, of immersion time – corroborated by EIS responses.

4.3.2 Wire Beam Electrode

The wire beam electrode (WBE) technique is a specialized method used for local electrochemical measurements and the evaluation of inhibitor performance on metallic surfaces. WBE involves the placement of a small-diameter wire onto the sample surface, creating a localized electrochemical cell. By applying a potential to the wire and measuring the resulting current, researchers can assess the local corrosion behavior and the impact of the imidazoline inhibitor in the immediate vicinity of the wire electrode. WBE data provide insights into localized corrosion processes, such as pitting and crevice corrosion, and can be used to evaluate the inhibitor's ability to suppress these detrimental processes. The interpretation of WBE measurements requires careful consideration of factors such as the wire electrode geometry, contact area, and the complex interplay between the local environment and the wire electrode surface. Developed to monitor localized corrosion, Zhao *et al.* used 0.5 mm wire beam electrodes in a 10×10 array, for which the wires were insulated 2 mm from each other using epoxy resin. TAI was used to evaluate the corrosion inhibition for N80 steel in CO_2-saturated NaCl-$NaNO_2$ solutions [Zhao, 2018].

The galvanic potentials and currents were mapped with the wires using a pre-programmed autoswitch device and ACM AutoZRA, with the galvanic potentials measured using SCE. The test was carried out for 10 hours, at 1-hour intervals, for which the galvanic currents and potentials were measured from 100 electrodes at room temperature. Shown in Figure 4.28, the potentials are mapped for test solutions having no inhibitor, and 200 mg/l TAI after 1 and 10 hours of exposure. The potentials became more positive with greater exposure time and in the presence of the inhibitor, and the currents became less anodic and more cathodic correspondingly in the current maps.

4.3.3 Atomic Force Microscopy

Atomic force microscopy (AFM) is a versatile characterization technique that enables the visualization of surface topography and the measurement of nanomechanical properties. By scanning a sharp probe over the sample surface, AFM produces high-resolution images, revealing details about the inhibitor-substrate interaction at the nanoscale. AFM can quantify surface roughness, measure surface forces, and provide information on adhesion forces between the inhibitor and the metal surface. This technique allows researchers to observe changes in surface morphology, assess the formation of protective films, and study the effects of inhibitors on surface properties. AFM is a surface-sensitive technique and may not capture the complete three-dimensional structure of the sample.

AFM was used in recent studies to analyze the effects of imidazoline inhibitors on the morphology of the steel surfaces after immersion in a corrosion medium. Singh *et al.* analyzed along with other spectroscopic techniques the surface of N80 steel after immersion in 15% HCl-containing solutions with AFM [Singh, 2020]. The 3D images in Figure 4.29 represent the surface without, and in the presence of chromeno pyrimidine (CPs), 8,8-Dimethyl-5-p-tolyl-8,9-dihyd ro-1H-chromeno (;)pyrimidine-2,4,6 (3H,5H,7H)-trione (CP-1) and 8,8-Dimeth

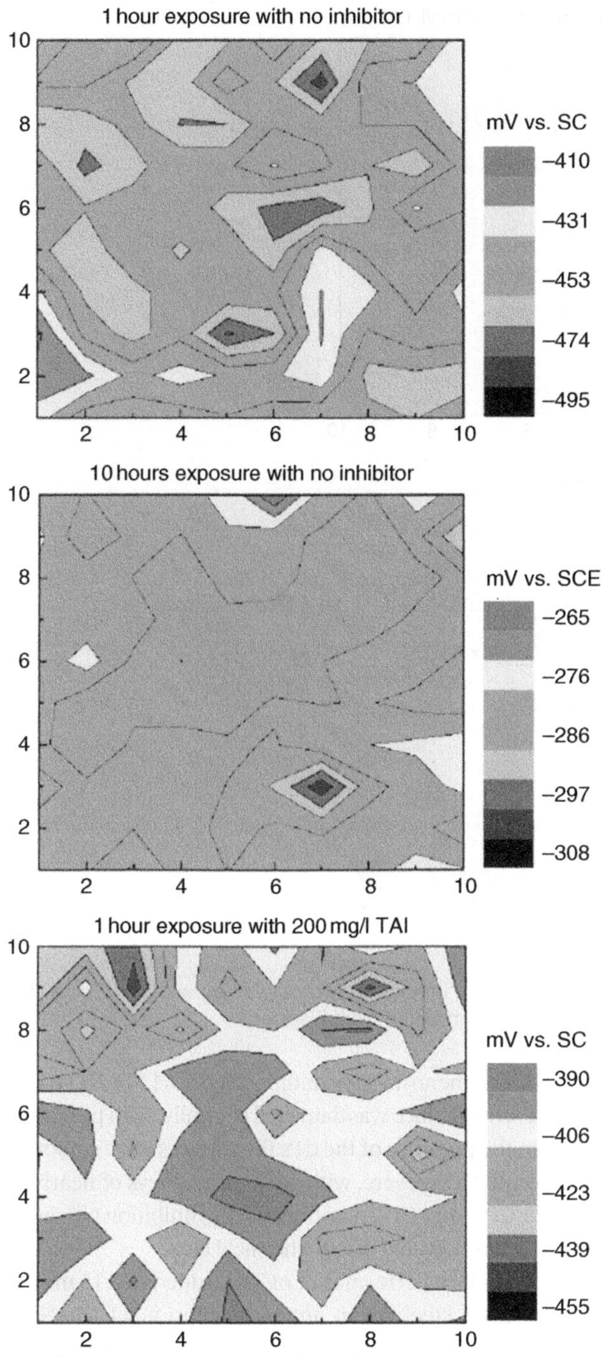

Figure 4.28 Potential maps measured on WBE exposed to the solutions containing no and 200 mg/L TAI at 1 and 10 hours. *Source:* Zhao [2018]/Elsevier/ CC BY 4.0.

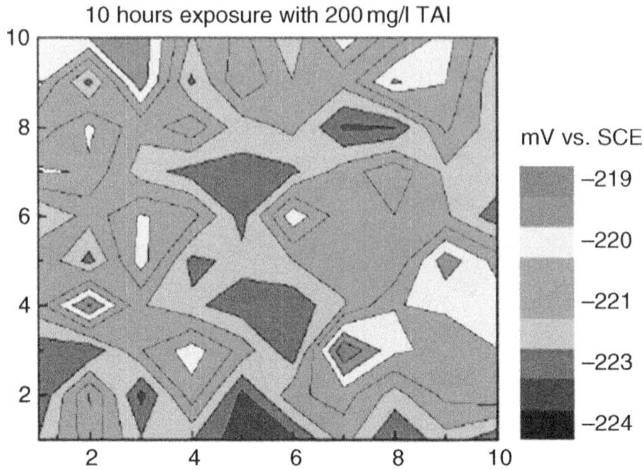

Figure 4.28 (Continued)

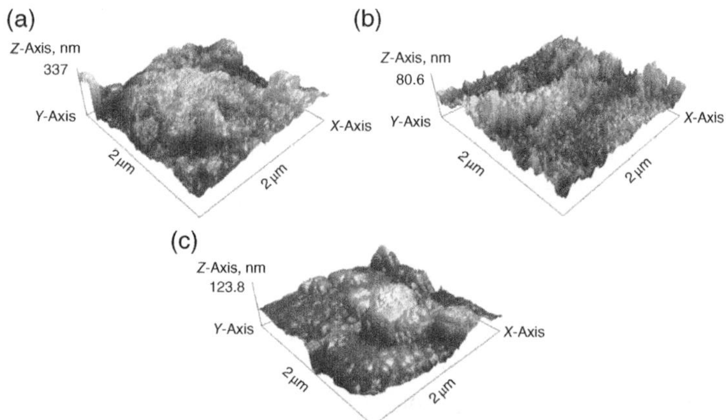

Figure 4.29 Three dimension AFM images (a) without CPs (b) with CP-1 (c) with CP-2. *Source:* Singh [2020]/with permission from Elsevier.

yl-5-phenyl-8,9-dihydro-1H-chromeno (;)pyrimidine-2,4,6 (3H,5H,7H)-trione (CP-2). In the absence of CPs, the surface was damaged severally, with peaks and valleys of around 330 nm. In the presence of the CPs the surfaces were smoother and the corrosion attack seemed less severe, with peaks and valleys of nearly 80 and 120 nm, respectively for CP-1 and CP-2, confirming the inhibition efficiency in order, accordingly with weight loss and electrochemical tests.

Wang *et al.* studied the inhibition performance of TAI, added in 0.15 mmol/dm^3, of the corrosion of Q235 steel in an aqueous solution of 2% NaCl saturated with CO_2, during two hours of immersion at 20 °C [Wang, 2011].

Adding the inhibitor showed that the surface remained flat and relatively uniform in topography. The steel surface roughness was nearly118 nm before immersion in the solution, increased to 450 nm after immersion in uninhibited solution because of corrosion that produced products, and decreased in the presence of TAI to 77 nm, indicating good inhibition efficiency. The authors quantified the adsorption of the film from force-distance curves from repeated experiments. In an uninhibited solution, the capillary force from water condensation between the tip and the sample was calculated on average as 2.66 nN. The adhesive force increased to 6.00 nN in inhibited solution, indicating that the adsorbed TAI film was hydrophobic, formed by the hydrophobic alkyl chains. The conclusions drawn considered the electrostatic repulsive force between the tip and sample surface at different displacements. Belarbi *et al.* carried out in-situ electrochemical/AFM investigations on the corrosion of API 5L X65 carbon steel in the absence and presence of 1-(2-aminoethyl)-2-oleyl-2-imidazolinium chloride [Belarbi, 2017]. The 3D in-situ AFM topography images were studied with corresponded surface profiles for samples in uninhibited solutions. The values of before-immersion samples were used as base values for analyzing the adsorption of the inhibitor. After immersion in the uninhibited solutions, the surface roughness increased with time to as high as 300 nm after three hours.

The values for solutions inhibited with 100 ppmv (part per million by volume) were studied. The maximum roughness after immersion for six hours was as high as 50 nm, attributed to effective adsorption of the inhibitor molecules onto the surface. The force-distance curves showed that the force between the tip and surface was zero when the distance was larger than 5 nm, indicating no interaction between the AFM tip and the surface. At 2 nm, the tip was attracted to the surface due to short-range attractive forces, at which the tip came in contact with the surface. In the presence of inhibitor film, it produced long-range repulsive intermolecular force pushing against the AFM tip. After six hours' immersion, the repulsive force reached maximum 0.5 nN.

Singh *et al.* examined three inhibitors AT-1, AT-2, and AT-3 on J55 steel [Singh, 2019]. They used NT-MDT multimode instrument for sample scanning of 10 mm × 10 mm. Without addition of an inhibitor the surface was heavily corroded of roughness of nearly 180 nm, and respectively for AT-1, AT-2, and AT-3, the roughness was nearly 10, 60, and 80 nm, indicating the superiority of AT-1. Okafor *et al.* evaluated the surface of N80 steel with AFM analysis, shown in Figure 4.30, in range of 0–2000 nm, after two hours of immersion in CO_2-saturated simulation formation solutions, with and without 100 ppm of RAIM at 40 °C [Okafor, 2011]. The protrusions and

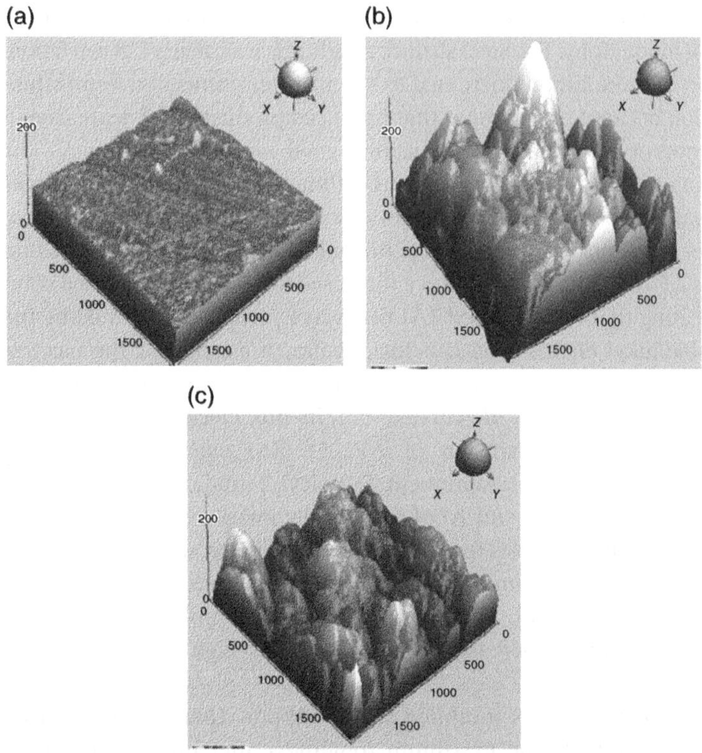

Figure 4.30 AFM three-dimensional images of (a) the unexposed (as-polished), in (b) CO_2-saturated simulated formation water solution, and (c) in (c) CO_2-saturated simulated formation water solution containing 100 ppm RAIM at 40 °C. Scale: X = nm; Y = nm; Z = nm. *Source:* Okafor [2011]/American Chemical Society.

surface roughness of nearly 25 nm made corrosion to occur at a high rate, to make roughness of nearly 360 nm in the uninhibited solutions. In the inhibited solutions, surface roughness was nearly 150 nm.

4.3.4 Scanning Tunnel Microscope

Scanning tunneling microscopy (STM) is a powerful imaging technique that provides atomic-scale insights into the surface morphology and electronic properties of imidazoline inhibitors. By scanning a sharp probe over the sample surface while maintaining a constant tunneling current, STM generates detailed topographic images. This technique allows for the visualization of individual atoms, surface defects, and the arrangement of molecules on the surface. STM can provide valuable information about the inhibitor's adsorption sites, surface coverage, and molecular arrangement. Additionally, STM-based spectroscopic techniques, such as scanning tunneling spectroscopy, can probe the electronic properties and energy levels of the inhibitor-substrate system. STM is limited to conductive samples and operates under ultra-high vacuum conditions. It was not common in studying the inhibitive effects of imidazolines in relation to time periods of immersion or other factors. Zhang *et al.* showed that 20 ppm of amide imidazoline formed a stable layer in CO_2-saturated solutions in a fashion nearly independent of immersion time with no potentials applied [Zhang, 2001], with images exhibiting no physical differences.

4.3.5 Polarization Modulation Infrared Reflection-Absorption Spectroscopy (PM-IRRAS)

Polarization modulation infrared reflection-absorption spectroscopy (PM-IRRAS) is a spectroscopic technique that measures the reflection and adsorption of infrared light to study molecular adsorption and conformation at the imidazoline inhibitor-metal interface. By analyzing the changes in the intensity and frequency of infrared light, PM-IRRAS provides information on the molecular structure, orientation, and interactions with the substrate. This technique can determine the adsorption modes of the inhibitor, investigate the formation of protective films, and monitor changes in the molecular conformation under different experimental conditions. PM-IRRAS data can be used to elucidate the bonding characteristics between the inhibitor and the metal surface and provide insights into the inhibitory mechanism. However, it should be noted that PM-IRRAS is surface-sensitive and requires the availability of suitable infrared-active vibrations. Desimone *et al.* used PM-IRRAS to evaluate the inhibition efficiency of AAOA, added in 0.82×10^{-5} M, in 5% NaCl CO_2-saturated solutions after two hours of immersion [Desimone, 2011]. The spectra of AAOA adsorption on steel surface are in Figure 4.31. The authors discerned the peaks of NH, CH_3, and carbonyl at between 3500 and 3100 cm^{-1}, at 2955 and 2872 cm^{-1}, and at 1636 cm^{-1}, respectively, and with no peaks for the C=N bond. They reported the symmetric and asymmetric stretching bands for CH_3 and CH_2 bands. Together with examining other peaks associated with amide bands and C—N and C—O stretching vibrators, the authors postulated that AAOA adsorption occurs by primary or secondary amine and/or amide functional groups. The authors evaluated also the molecular orientation of the adsorbed molecules and explained the physicochemical nature of the adsorbed film. From the wave number of CH_2 the authors indicated that the AAOA

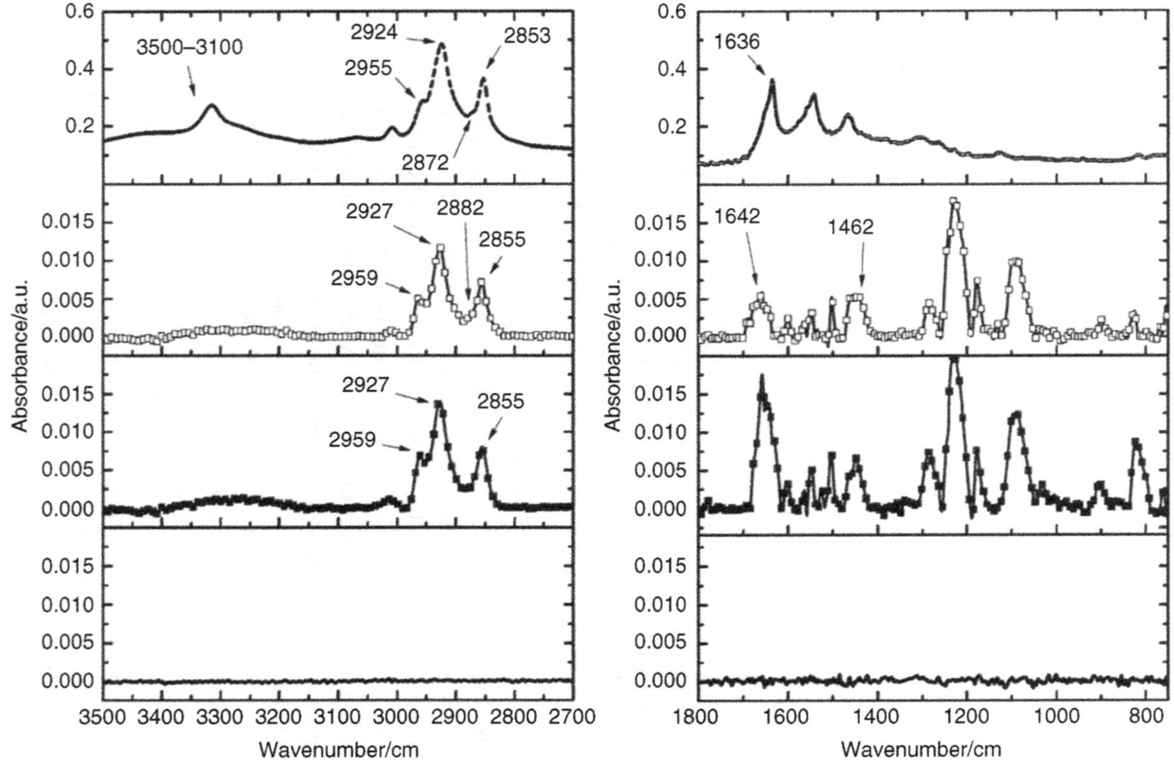

Figure 4.31 PM-IRRAS spectra of the corrosion inhibitor on carbon steel sample collected after the electrode was immersed two hours at E_{corr} in an aqueous inhibited solution containing 0.82×10^{-5} M of the inhibitor and 5% NaCl saturated with CO_2 at 25 °C: ((–) before immersion (blank), (Black Square) spectrum obtained after the immersion time, (White Square) spectrum obtained after the washing), and IR spectrum of a KBr disc of AAOA after the synthesis (--- (bottom figures)). *Source:* Adapted from Desimone [2011].

monolayer has an amorphous, disordered liquid-like structure – in comparison to previous investigations on ordered monolayers – attributing that to surface roughness and surface dissolution.

4.3.6 Energy-dispersive X-ray Spectroscopy

Energy-dispersive X-ray spectroscopy (EDX) is an elemental analysis technique that allows for the identification and mapping of elements present in imidazoline inhibitors in oil flow environments. By detecting characteristic X-rays emitted from the sample surface upon exposure to an electron beam, EDX provides information on the elemental composition and distribution. This technique can determine the presence and concentration of elements, allowing researchers to assess the spatial distribution of inhibitors and correlate it with corrosion inhibition performance. EDX data can be used to study the penetration and distribution of inhibitors on the metal surface and evaluate their effectiveness in forming protective layers. However, EDX has limitations in terms of spatial resolution and may not provide information about the chemical state of elements.

Ding *et al.* analyzed the corrosion products and their composition in exposure to CO_2-saturated autoclave environments in the absence and presence of DETA/TOFA imidazoline at high temperatures [Ding, 2018]. EDS showed the distribution of iron, carbon, and oxygen, indicating, together with XRD, the formation of iron oxides, iron carbide, and iron carbonate.

4.3.7 X-Ray Diffraction

X-ray diffraction (XRD) is a powerful technique used to analyze the crystal structure, phase identification, and corrosion product characterization of imidazoline inhibitors on pipeline steels. By directing X-rays at the sample, XRD measures the diffraction pattern produced by the crystal lattice, providing information about the arrangement of atoms and crystallographic phases present. XRD can identify the presence of corrosion products, detect changes in crystal structure due to inhibitor interaction, and assess the degree of crystallinity. This technique allows researchers to gain insights into the inhibitory mechanisms, identify the formation of protective films, and study the crystallographic changes induced by the inhibitors. XRD requires powdered samples or single crystals and it may not provide information on the amorphous or non-crystalline phases.

XRD was used in some studies for which high amounts of corrosion products formed in autoclave environments. Ding *et al.* used XRD to analyze the corrosion products of API 5L X65 steel in autoclave CO_2-saturated environments in the absence and presence of DETA/TOFA imidazoline, of 440 ppm at 120 °C and 880 ppm at 150 °C in single- and double-autoclave systems [Ding, 2018]. No XRD peaks appeared for iron carbonate for corrosion products in the presence of inhibitor at 120 °C. At 150 °C, peaks for Fe_3O_4 appeared even in the presence of the inhibitor, but it was less in intensity than that in the uninhibited solution, suggesting that the thickness of the corrosion products was accordingly less.

In a double-autoclave system, in the presence and absence of the inhibitor at 120 °C, no XRD peaks for $FeCO_3$ appeared, due to the relatively shorter time of exposure, during which less interfacial concentrations for Fe^{2+} accumulated for iron carbonate precipitation. At 150 °C, the XRD data were similar to those of the single-autoclave system, both for the uninhibited and inhibited environments. In the inhibited solutions, the magnetite Fe_3O_4 peaks were less in intensity than that of the uninhibited solution, suggesting thinner corrosion products layer formed.

Wang *et al.* studied the effects of imidazoline (IM), 1-butyl-2-thiobenzylbe nzimidazole (BTBBI) and lauric acid (LA) on N80 carbon steel of galvanic corrosion with 13Cr stainless steel in supercritical-CO_2 oilfield-produced water [Wang, 2019]. The test solutions contained 62.36 g/l NaCl, 3.4 g/l KCl, 0.57 g/l $CaCl_2$, 0.64 g/l Na_2SO_4, 4.46 g/l $MgCl_2$•$6H_2O$, 0.52 g/l $NaHCO_3$, and the concentrations of the inhibitors were 100 mg/l, purged with CO_2 (99.99%) for 12 hours, and the pH was 5.3. Imidazoline exhibited the best performance; it showed that $FeCO_3$ and F_3C appear persistently regardless of if the two metals coupled or not, with minor differences in relation to the intensity peak angles. In the presence of the inhibitors, the strong signal of iron substrate might have concealed the peaks of the corrosion products, or the corrosion products might not have been crystalline. Xiang *et al.* used XRD to evaluate N80 steel samples exposed to supercritical CO_2 SC-CO_2 phase (SP) and SC-CO_2-saturated aqueous phase (AP), containing SO_2, NO_2, and O_2 impurities, in the presence and absence of imidazoline and piperazine inhibitors, at 10 MPa and 60 °C, for 24 hours [Xiang, 2017]. The concentrations were 300 and 1000 ppm, and the corrosion layer seemed to be thin, as iron was detected in all samples. In the supercritical CO_2 conditions (SC) samples, mikasaite ($Fe_2(SO_4)_3$) was detected while it was not detected in the aqueous phase (AP) condition samples, and iron carbonate ($FeCO_3$) was not detected in all samples. The stronger iron peaks of AP samples was attributed to greater products thickness than those of SC samples.

4.3.8 Fourier-Transform Infrared Spectroscopy

Fourier-transform infrared spectroscopy (FTIR) is a spectroscopic technique used to analyze the vibrational modes and molecular composition of imidazoline inhibitors in oil flow environments. By measuring the absorption and transmission of infrared light, FTIR provides information about the functional groups and chemical bonds present in the inhibitor. This technique allows researchers to identify specific functional groups associated with inhibition, study the adsorption behavior of the inhibitor on the metal surface, and monitor changes in the molecular structure under different conditions. FTIR data can be used to infer the nature of inhibitor-substrate interactions, assess the formation of protective layers, and investigate the impact of environmental factors on inhibitor performance. FTIR requires sample preparation, and it is sensitive to the presence of other compounds or impurities that may interfere with the spectra.

Li *et al.* used FTIR to analyze the adsorption of 0.1 g/l of 1-(2-aminoethyl)-2-heptadecyl imidazoline on N80 carbon steel in CO_2-saturated 1.65 wt% NaCl solutions containing 1 and 3 g/l HAc [Li, 2017]. The analysis was for corrosion inside and outside of the artificial crevice of 12 mm × 12 mm with a thickness of 300 µm. The authors studied the transmittance spectra for samples in an uninhibited solution, inhibitor added in the beginning, and after 24 hours of the experiments. Obviously, there were no absorption peaks for the sample in the uninhibited solution, and for the inhibitor added in the beginning of the test, there was an absorption peak at about 1625 cm^{-1} for the outside crevice surface and the peak for the inside crevice surface was less. With adding the inhibitor after 24 hours, the absorption peak of the inhibitor was observed outside the crevice, with no peaks for inside the crevice, as the corrosion products limited the diffusion of the inhibitor molecules. The characteristic absorption peaks correspond to C=N in the inhibitor, in reference to the previous studies [Yoo, 2012, Xiong, 2016]. Similar absorption peaks were reported by Chen *et al.* at 1608 cm^{-1}, attributed to C=N in the ring of imidazoline inhibitor synthesized from oleic acid and diethylenetriamine [Chen, 2015].

The composition of imidazoline inhibitor was verified by Chen *et al.* using FTIR, shown in Figure 4.32, and the strong absorption peak was evident at 1608 cm^{-1} of IR, attributed to the C—N bond in the IM ring [Chen, 2015].

Migahed *et al.* used a FTIR spectrophotometer and Varian Mercury VXR-300 NMR spectrometer (for which dimethyl sulfoxide (DMSO-d_6) was used as solvent) to analyze the infrared (IR) spectra and ^1H nuclear magnetic resonance (NMR) spectra, respectively, for the synthesized PQI and SQI [Migahed, 2017]. The absorption bands for PQI compound were C—H, CH_3 at 2851, 2922 cm^{-1}, respectively, $(CH_2)_n$ skeletal at 723 cm^{-1}, primary amine at 3301 cm^{-1}, and C=N had strong IR band at 1642 cm^{-1}, and the imidazoline ring had a relatively weak IR band at 1603 cm^{-1}. The absorption bands for SQI compound were C—H, CH_3 at 2847, 2919 cm^{-1}, $(CH_2)_n$ skeletal at 722 cm^{-1}, primary amine at 3293 cm^{-1}, and C=N had strong IR band at 1642 cm^{-1}, and the imidazoline ring had a relatively weak IR band at 1604 cm^{-1}.

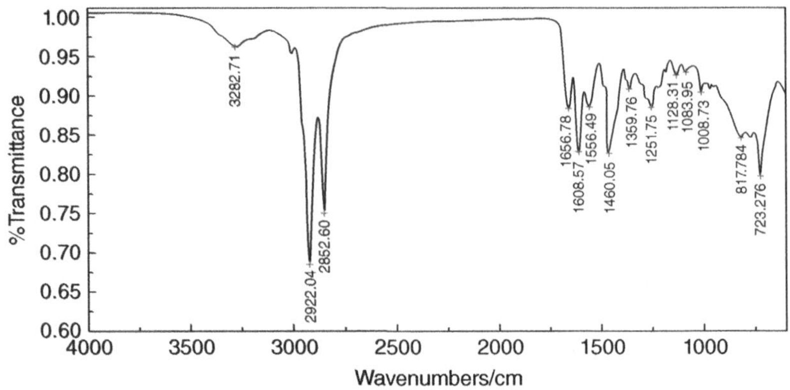

Figure 4.32 FTIR spectra of IM. *Source:* Chen [2015]/Emerald Group Publishing Limited.

The structure of the synthesized PQI and SQI compounds were verified using ^{1}HNMR spectrum, depicted in the study. For PQI compound, the NMR chemical shift for CH_3 was at 0.837, 0.834, 0.857 ppm, for $(CH_2)_n$ at 1.242 ppm, for CH_2 attached to nitrogen of imidazoline ring at 2.497, 2.501, 2.503 ppm, for CH_2 attached to carbon of imidazoline ring at 2.207, 2.222, 2.234 ppm, for equivalent methylene group in imidazoline ring at 3.591, 3.613 ppm. For SQI compound; the chemical shift for CH_3 was at 0.857, 0.844, 0.837 ppm, for $CH_2)_n$ at 1.243 ppm, for CH_2 attached to nitrogen of imidazoline ring at 2.494, 2.500, 2.506 ppm, for CH_2 attached to carbon of imidazoline ring at 2.210, 2.234, 2.258 ppm, for equivalent methylene group in imidazoline ring at 3.586, 3.608 ppm, for H of NH_2 group at 7.104 ppm.

4.4 Weight Measurements

4.4.1 Quartz Crystal Measurements

Quartz crystal measurements (QCM) would provide in some cases valuable insights into the mass changes occurring on the surface of pipeline steels in the presence of imidazoline inhibitors. This technique involves coating a quartz crystal resonator with the metal specimen and monitoring the changes in its resonant frequency. The frequency shift is directly proportional to the mass change on the crystal surface, allowing for quantitative measurements of corrosion rates. By conducting QCMs under controlled experimental conditions, researchers can assess the effectiveness of imidazoline inhibitors in reducing the corrosion rate of pipeline steels. The data obtained from QCMs can be used to calculate corrosion rates, evaluate the inhibitory efficiency of the imidazoline inhibitor, and compare the performance of different inhibitors. Interesting experimental findings may include significant reductions in the corrosion rate with the addition of the imidazoline inhibitor, providing evidence of its effectiveness in protecting pipeline steels in oil flow environments. However, it is important to consider the limitations of QCMs, such as the assumption of uniform corrosion and the influence of surface roughness on frequency measurements.

The weight difference that resulted from adsorption of inhibitor film of TOFA/DETA imidazoline was studied by Jevremović *et al.*, after immersion in CO_2-saturated 3 wt% NaCl solutions for one hour [Jevremović, 2016]. The analysis was made on an iron-coated quartz crystal to measure the weight difference by corrosion and the adsorption of the inhibitor molecules [Yurt, 2011]. QCM technique is based on the relationship between the mass and frequency change of time, governed by Sauerbrey equation as:

$$\Delta f = -\frac{2f_o^2}{\sqrt{\mu_q \rho_q}} \Delta m \tag{4.12}$$

where Δf is the resonant frequency shift, Δm is the change in mass per unit area, f_o is the fundamental frequency of the resonator, ρ_q is the density of the quartz crystal, and μ_q is the shear modulus of the quartz crystal [Binks, 1999]. Without inhibitor, the mass decreased by 180 μg/cm^2 as a result of iron dissolution, as revealed from nearly linear relationship between the mass loss and time. With the addition of 90 ppm TOFA/DETA imidazoline, the mass loss showed a decrease after 20 minutes. The initial mass loss was attributed to replacement of water molecules by the inhibitor molecules, in reference to analysis elsewhere [Kern, 2002].

4.4.2 Weight Loss Measurements

Weight loss measurements are a commonly used technique to evaluate the corrosion behavior of pipeline steels in the presence of imidazoline inhibitors. This method involves exposing metal specimens to the corrosive environment, either with or without the inhibitor, and monitoring the loss of weight over a specific period. The weight loss is an indicator of the corrosion rate, allowing researchers to assess the effectiveness of the imidazoline inhibitor in slowing down corrosion. By comparing weight loss data from different experiments, researchers can quantify the inhibitory performance of the imidazoline inhibitor and analyze the corrosion mechanism. The experimental findings may include a substantial reduction in weight loss in the presence of the imidazoline inhibitor, indicating its ability to protect pipeline steels against corrosion in oil flow environments. Weight loss measurements require careful sample preparation, potential errors due to measurement uncertainties, and the inability to provide real-time data on the corrosion process. Additionally, weight loss measurements may not provide detailed information on localized corrosion phenomena or the formation of protective films on the metal surface. Singh *et al.* evaluated with weight loss experiments the inhibition efficiency of three inhibitors AT-1, AT-2, and AT-3 on J55 steel [Singh, 2019]. The inhibition efficiency increased with the concentration, attributed to a greater surface area covered by the inhibitor molecules accordingly. Langmuir isotherm showed a good fit for the adsorption for all inhibitors, represented as:

$$\frac{C_{\text{inh}}}{\theta} = \frac{1}{K_{\text{ads}}} + C_{\text{inh}} \tag{4.13}$$

where C is the inhibitor concentration in (mg/l), θ is the surface coverage, and K_{ads} is the equilibrium interfacial constant for the inhibitor adsorption/desorption. The negative values of $\Delta G^{\text{o}}_{\text{ads}}$, between −20 and −40 kJ/mol, indicated spontaneous, physicochemical adsorption.

Li *et al.* investigated with loss measurements the inhibition of 1-(2-aminoethyl)-2-heptadecyl imidazoline, added with 0.1 g/l IM, on the corrosion and weight loss of N80 carbon steel in CO_2-saturated 1.65 wt% NaCl solutions containing HAc for 72 hours [Li, 2017]. The measurements were for an artificial crevice system for which specimens as working electrodes inside and outside the crevice were connected during weight loss tests. The authors evaluated the inhibitor added in the beginning and after 24 hours of the experiment in solutions containing 1 and 3 g/l HAc. Adding the inhibitor in the beginning reduced the corrosion rates nearly 10 times more than when added after 24 hours and weight loss was higher in the 3 g/l HAc solutions. And, expectedly, the corrosion rates in the crevice were higher than outside the crevice.

Zhang *et al.* evaluated the significance of thioureido imidazoline on the weight loss of X70 steel immersed with different surface roughness values of 150- and 5000-grit in CO_2-saturated solutions [Zhang, 2019]. As shown in Figure 4.33, the corrosion rates, measured in mm/y, were corresponded to the surface condition and immersion periods of 1, 24, and 240 hours in the presence and absence of 1.22×10^{-5} mol/l. The inhibitor efficiency with longer periods of immersion was higher and the corrosion rates were generally less for 5000-grit surfaces.

In dynamic supercritical CO_2 oilfield water conditions, Wang *et al.* studied the weight loss performance of N80 carbon steel coupled with 13Cr stainless steel in the presence of IM, BTBBI and LA, with durations of

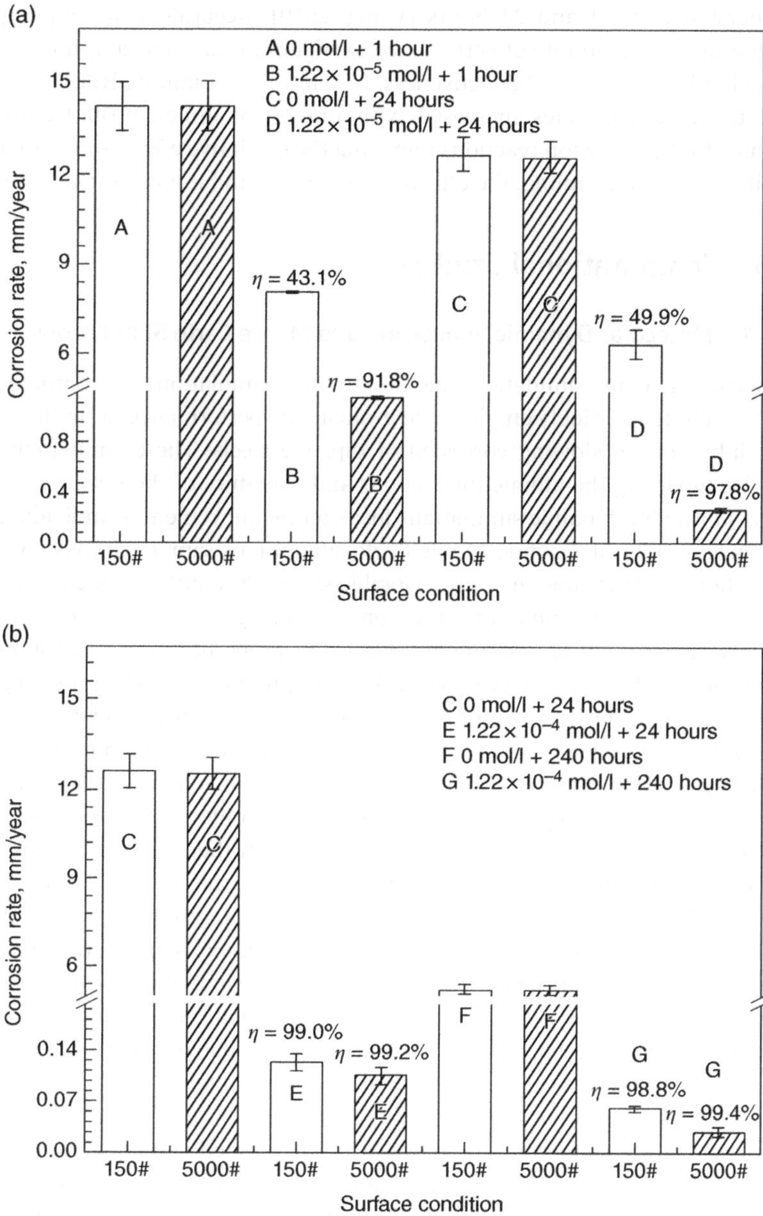

Figure 4.33 General corrosion rates and inhibition efficiencies (obtained from weight loss measurements) of samples with different surface roughness after immersion for different time in 3 wt% NaCl solution with various concentrations of inhibitor at 333 K. (a) A: uninhibited, 1 hr, B: 1.22x10-5 mol/l, 1 hr, C: Uninhibited, 24 hr, D: 1.22x10-5 mol/l, 24 hr, (b) C: Uninhibited, 24 hr, E: 1.22x10-4 mol/l, F:24 hr, F: Uninhibited, 240 hr, G: 1.22x10-4 mol/l, 240 hr. *Source:* Zhang [2018]/with permission from Elsevier.

immersion being 4 and 24 hours [Wang, 2019]. Regardless of coupling (when the corrosion rate of carbon steel is higher by galvanic coupling) or not, all inhibitors showed effectiveness. In solutions containing BTBBI and LA, the corrosion rates increased with immersion time, attributed by authors to the corrosion reactions impeding the inhibitive effects of the two inhibitors. Upon coupling, the effectiveness of the inhibitors was less.

4.5 Computational Studies

4.5.1 Molecular Dynamic Simulation and Monte Carlo Simulations

Molecular dynamic simulations and Monte Carlo simulations are computational techniques widely employed to evaluate the performance of imidazoline inhibitors in slowing corrosion of pipeline steels. These simulations involve modeling the atomic interactions and dynamics of the system over time. Molecular dynamic simulations focus on the movement of individual atoms and molecules, while Monte Carlo simulations sample various configurations to determine the most favorable states. By simulating the behavior of the imidazoline inhibitor at the atomic scale, these techniques generate valuable data on structural changes, diffusion properties, and adsorption behavior. The data obtained can be used to interpret the results by providing insights into the mechanisms of inhibitor-metal interactions, understanding the inhibition efficiency, and predicting the performance of different inhibitor designs. Major findings from these simulations include the determination of adsorption energies, exploration of inhibitor conformational changes, and investigation of the inhibitor's ability to disrupt the corrosion process. However, the simulations have limitations, such as the need for accurate force field parameters, the challenge of accurately representing the complex dynamics of the system, and the reliance on assumptions and approximations. Despite these limitations, molecular dynamic simulations and Monte Carlo simulations are powerful tools in computational evaluation, complementing experimental approaches and providing valuable insights into the corrosion inhibition mechanisms of imidazoline inhibitors.

MD simulations are useful for prediction of the molecular properties and possible reactive sites in the presence of reactive functional groups, heteroatoms, and the inhibitor molecules. Monte Carlo techniques were used to investigate in deterministic simulations the reliability of oil pipelines against corrosion and later were used to investigate adsorption of inhibitors [Zhang, 2012, Verma, 2018, Sukul, 2018, Tantawy, 2020]. Tantawy *et al.* used Metropolis Monte Carlo simulation to study the adsorption of PGS-8, PGS-10, and PGS-12 inhibitors of chemical composition mentioned in earlier sections.

The authors depicted the top and side views of the steadiest adsorption configurations of PGS-8, PGS-10, and PGS-12 on Fe (110). They showed that the moiety of inhibitor molecules could interact with the Fe interface. 1,3 benzodioxolane takes near-flat orientation while alkyl group takes perpendicular direction on the Fe surface. The order of adsorption energies of PGS-8, PGS-10, and PGS-12 on Fe (110) surface showed agreement with the inhibition efficiency, investigated from an electrochemical method.

Singh *et al.* carried out molecular dynamic simulations of the interactions between CP, CP-1 and CP-2 and the N80 steel surface [Singh, 2020], of top and side views, representing the most stable configuration of adsorption. CP-1 showed a flat orientation (and thus higher efficiency of greater surface coverage), and CP-2, due to the phenyl ring, was perpendicular to the metal surface. CP-1's calculated binding energy was nearly 830 kJ/mol and that of CP-2 was nearly 810 kJ/mol, suggesting higher adsorption, in line with a previous finding [ElBelghiti, 2016].

Singh *et al.* carried out MD simulation, with BIOVIA Materials Studio® commercial software, for the inhibitor AT-1 [Singh, 2019]. Fe (110) surface with a slab of 5 Å with a simulation box of $24.82 \times 24.82 \times 35.69\,\text{Å}^3$, containing one inhibitor molecule, $9Cl^-$, $491H_2O$, and $9H_3O^+$. The strength of adsorption was estimated with the following equation [Lgaz, 2017]:

$$E_{\text{interaction}} = E_{\text{total}} - \left(E_{\text{surface+solution}} + E_{\text{inhibitor}}\right) \tag{4.14}$$

E_{total}, $E_{\text{surface+solution}}$, and $E_{\text{inhibitor}}$ represent the system energy, the energy of Fe (110) and corrosive medium without inhibitor molecules, the energy of inhibitor molecules, respectively. The simulations were run for equilibrium conditions for which the inhibitors adsorb horizontally with strong interactions sustained by donation, back-donation, and acceptance of electrons through π electrons of the five member and phenyl rings, and via the lone pair of the heteroatoms, for neutral inhibitors and protonated inhibitors, respectively. AT-3 inhibitor molecule was observed as not too far from the surface, suggesting a lesser role for electron withdrawing nature of the nitro group. The high interaction energies indicated the strong adsorption, with more than one bond per inhibitor, in correlation to the number of methoxy groups (electro-donating groups). The decrease in the interaction energy of protonated inhibitors is due to their strong solvation energy, reducing their adsorption.

4.5.2 Density Functional Theory

Density functional theory (DFT) is another computational method used to study the imidazoline inhibitor's interactions with pipeline steels at the electronic level. DFT calculations involve solving the quantum mechanical equations that describe the system to obtain electronic properties and energies. By employing DFT, researchers can analyze the electronic structure, charge transfer, and reactivity of the inhibitor-metal interface. This data enables the interpretation of the results by providing a deeper understanding of the inhibitor's bonding mechanism, electronic properties, and their influence on corrosion inhibition. For instance, DFT calculations can reveal the changes in electronic density distribution upon inhibitor adsorption, identify the sites of highest reactivity, and quantify the thermodynamic stability of the inhibitor-metal complex. The limitations of DFT are linked to the approximation of exchange-correlation functionals and the computational cost associated with large systems. Despite these limitations, DFT offers valuable insights into the electronic properties and bonding behavior of imidazoline inhibitors, enhancing the understanding of their corrosion inhibition mechanisms.

Quantum chemical calculations are utilized to determine the molecular structures of the imidazoline inhibitors and their ability to inhibit reactions. Energy of highest occupied molecular orbital (E_{HOMO}), energy of lowest unoccupied molecular orbital (E_{LUMO}), energy difference ($\Delta E_{L\text{-}H}$), involving (μ_d), (χ), (I), (A), (η_G), (σ_S), (ΔN), corresponding to dipole moment, electronegativity, ionization potential, electron affinity, global hardness, softness and number of transferred electrons from an inhibitor molecule to the carbon steel, respectively, are the quantum parameters used in that respect. The interfacial interactions between the metal surface and inhibitor molecules depend on the distribution of frontier molecular orbitals, HOMO and LUMO energies. HOMO energy represents the ability of electron donation, and LUMO energy represents the ability of electron acceptation. The higher the HOMO energy and the lower the LUMO energy, the higher the tendency for the electron donation and electron acceptance for the inhibitor molecules, respectively. The reaction of reactants depends on HOMO and LUMO energies, which are associated with the ionization potential values. The energy difference ($\Delta E_{L\text{-}H}$) indicates the stability of the inhibition; the lower the $\Delta E_{L\text{-}H}$, the more stable the interaction between the inhibitor molecules and the surface, thus, it is of higher inhibition efficiency.

Tantawy *et al.* analyzed the optimized molecular structure of inhibitors, HOMO and LUMO frontier molecular orbitals, and molecular electrostatic potential (MEP), for PGS-8, PGS-10, and PGS-12, shown in Figure 4.34 [Tantawy, 2020].

The authors concluded that inhibitors of high E_{HOMO} values have good inhibition efficiency, as E_{HOMO} indicated the electron donation tendency to species of low hollow molecular orbital energy [Musa, 2010]. E_{LUMO} indicates the ability of a molecule to accept an electron and, therefore, the lower the E_{LUMO}, the easier is the acceptance of electrons from the metal d-orbitals

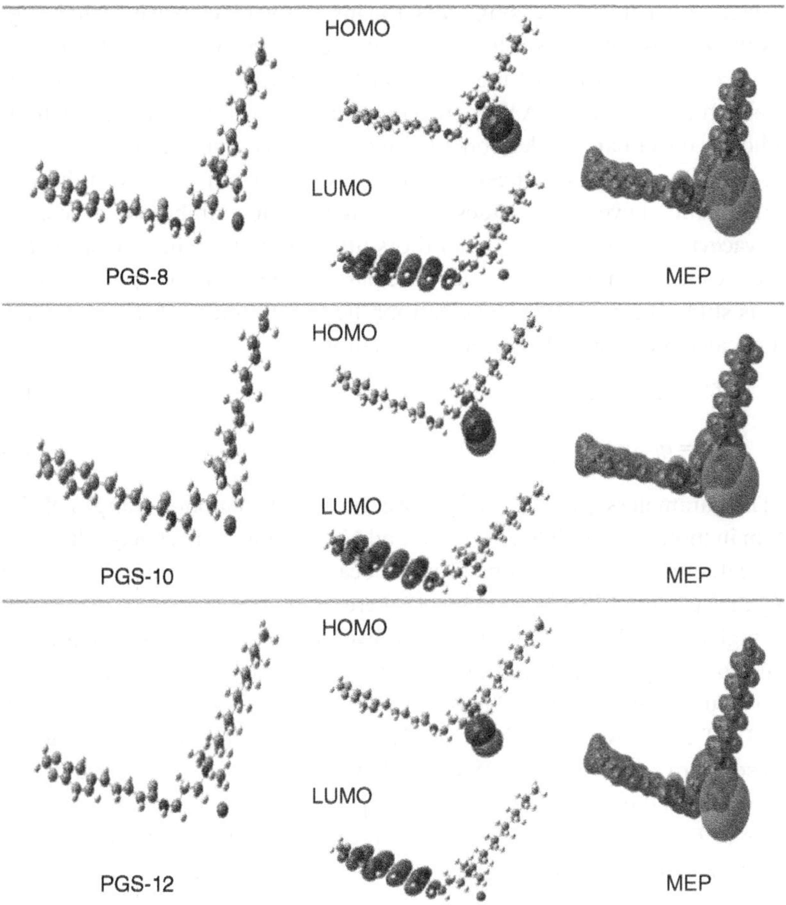

Figure 4.34 Fully optimized, HOMO and LUMO and the molecular electrostatic potential of the studied PGS inhibitors. *Source:* Tantawy [2020]/with permission of Elsevier.

[Finšgar, 2008]. The inhibitors in that investigated case were suggested to accept electrons from the d-orbital of a metal and to give electrons to unoccupied d-orbitals. The trends of values of E_{HOMO} and E_{LUMO}, of densities located on iodide ion, and on 1,3 benzodioxolane and C═O group, respectively, agree empirically with the electrochemical findings. MEPs showed the low and high electron density regions in the gas phase, in the study of red color of most negative potentials (associated in majority with the iodide ion and C═O group) and blue color of positive potentials. The energy gap (ΔE), tabulated, accounts for the reactivity of the inhibitors with the steel surface which, if in decreasing order, the inhibition efficiency is higher, suggesting in this case PGS-8 has the lowest reactivity, which had high electronegativity, readily reaching equalization [Geerlings, 2002]. The fraction of electron transferred (ΔN) increasing from lower to higher chain length indicates the enhanced electron donation between the studied inhibitor and the surface. Global hardness (η) indicates the stability of the inhibitor which, the lower its values, the higher the inhibition efficiency [Kovačević, 2011]. Besides, the authors utilized Fukui indices to study the reactive electrophilic and nucleophilic parts of the inhibitors based on previous study [Lgaz, 2017]. The electrophilic f^- and nucleophilic f^+ centers were calculated using the following equations:

$$f^+ = q_{(N+1)} - q_{(N)} \tag{4.15}$$

$$f^- = q_{(N)} - q_{(N-1)} \tag{4.16}$$

The parameters $q_{(N+1)}$, $q_{(N-1)}$, $q_{(N)}$ are the Mulliken atomic charges of the atom in molecules with $N+1$, $N-1$, and N electrons, respectively. The positive extent of Fukui functions f^+ were located on C5, C7, and C27 and the negative extent of Fukui functions f^- were located on I.

Singh *et al.* calculated at DFT/GGA level the optimization of the ground states of the synthesized inhibitors Becke One Parameter (BOP) functional and Double Numeric plus Polarization (DNP) basis set for three inhibitors AT-1, AT-2, and AT-3 [Singh, 2019]. The effect of solvation effects (aqueous phase) were controlled by COSMO and the calculations were carried out by Materials Studio [Materials Studio, 2013]. The authors outlined the correlations of E_{HOMO} and E_{LUMO} to ionization energy (I), electronic affinity (A), electronegativity (χ) and hardness (η) as [Dewar, 1977, Pearson, 1990]:

$$I = -E_{\text{HOMO}} \tag{4.17}$$

$$A = -E_{\text{LUMO}} \tag{4.18}$$

$$\chi = -\frac{1}{2}\left(E_{\text{HOMO}} + E_{\text{LUMO}}\right) \tag{4.19}$$

$$\eta = -\frac{1}{2}\left(E_{\text{HOMO}} - E_{\text{LUMO}}\right) \tag{4.20}$$

The optimized structures along with HOMO and LUMO iso surfaces for protonated AT-1, AT-2, and AT-3 inhibitors are shown in Figure 4.35. The smaller the energy gap between the inhibitor molecules and iron cluster, the stronger the interaction and adsorption. From the data, the authors reported that the electron donating OCH3 (AT-1) functional group and parent (AT-2) inhibitors donate electrons to the metal surface, while electron withdrawing

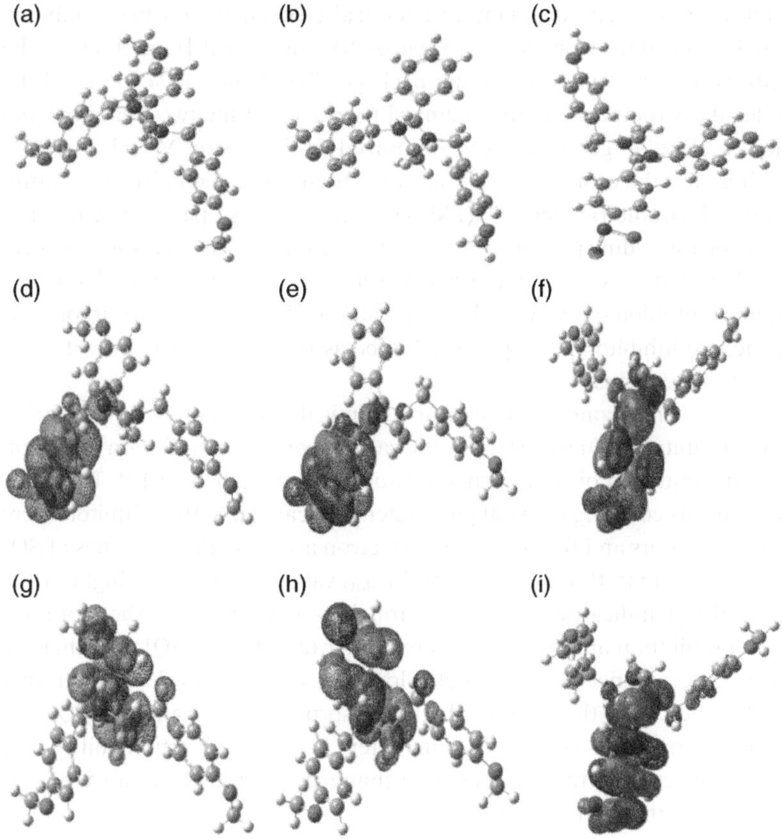

Figure 4.35 (a–c) Optimized geometries of protonated inhibitors (a) AT-1 (b) AT-2 (c) AT-3. (d–i) Frontier molecular orbital's of protonated AT-1 (d) HOMO (g) LUMO, AT-2 (e) HOMO (h) LUMO and AT-3 (f) HOMO (i) LUMO. *Source:* Singh [2020]/with permission from Elsevier.

NO_2 (AT-3) functional group inhibitor would accept the electron from metal through back-donation. In addition, they reported that the protonated inhibitors would take electrons from the metal surface through back-bonding, concluding that the electron donating group donate the electrons to the metal in neutral form and accept the electrons by back-donation in protonated form, and the electron withdrawing group accept the electrons by back-donation both in neutral and protonated forms. For AT-1 and AT-2, HOMO was distributed over the imidazoline ring and oxygen (O_{47}) and for AT-3, HOMO was distributed over the phenyl ring. LUMO was distributed over the phenyl ring for all inhibitors but also over NO_2 for AT-3.

Liu *et al.* carried out quantum chemical calculations on two inhibitors AEI-11 and AQI-11 on the corrosion of N80 mild steel [Liu, 2009A]. The authors used molecular modeling package ChemOffice 2004 to model the molecular structures and the chemical properties of the two inhibitors and their molecular geometry was optimized with Austin Model 1 (AM1) method. For the orbital calculations, they used the semiempirical quantum chemical parametric method 3(PM3) model in MOPAC package, with tabulated corresponding parameters in the study. E_{HOMO} of AQI-11 was less negative than that of AEI-11, and it had lower energy gap indicating that it had higher inhibition efficiency. The long chain hydrophobic hydrocarbon tails of the two inhibitors were proposed to be associated with the formation of the protective films.

Migahed *et al.* conducted quantum chemical calculations for two synthesized inhibitors, PQI and SQI, for the sake of corrosion inhibition of carbon steel in simulated oil well formation solutions [Migahed, 2017]. The tabulated results covering relevant parameters indicated that the inhibitors were electron donors and the surface was electron acceptor. E_{HOMO} values of SQI were higher than those of PQI, and E_{LUMO} values of PQI were higher than those of SQI, indicating that the electron donation ability, and therefore corrosion inhibition ability of SQI, were higher than that of PQI, and that the electron acceptance ability of SQI is low. $\Delta E_{L\text{-}H}$ of SQI was lower than that of PQI, indicating that SQI had higher adsorptive ability, and μ of SQI was higher than that of PQI, indicating better corrosion inhibition ability. The authors calculated the number of exchanged electrons from an inhibitor molecule to the steel surface as:

$$\Delta N = \frac{\chi_{Fe} - \chi_{inh}}{2(\eta_{GFe} + \eta_{inh})} \tag{4.21}$$

where (χ_{Fe}) is the electronegativity of iron (which is 7 eV/mol), (χ_{inh}) is the electronegativity of a synthesized inhibitor molecule, (η_{GFe}) is the hardness of iron (which is 0 eV/mol), and (η_{inh}) is the hardness of the inhibitor molecule. The calculated ΔN values were under 3.6, indicating that the inhibition is proportional with electron donation ability at the surface. ΔN for SQI was higher than that of PQI, indicating higher electron donation ability and inhibition efficiency. Frontier molecule orbital density distributions of PQI and SQI compounds were depicted in the study.

4.5.3 Radial Distribution Function

Radial distribution function (RDF) analysis is a computational technique used to investigate the structural arrangement of particles in a system. By analyzing the RDF, researchers can determine the probability distribution of distances between atoms or molecules, providing insights into the local ordering and coordination of species. In the context of imidazoline inhibitors, RDF analysis can shed light on the spatial arrangement of inhibitor molecules around the metal surface, offering valuable information about the formation of protective layers or adsorption behavior. The data can be used to interpret the results by assessing the degree of ordering, examining the inhibitor's ability to form a well-defined interfacial structure, and understanding the influence of inhibitor concentration or temperature on the arrangement of inhibitor molecules. Interesting findings from RDF analysis include the identification of preferred adsorption sites, the characterization of surface coverage, and the evaluation of inhibitor clustering phenomena. RDF analysis is limited to capturing only spatial information and does not provide information on dynamic processes or time-dependent behavior.

RDF analysis gives better details on the interactions between metal and inhibitor molecules and was used for structural characterization of materials. RDFs were based on simulation trajectories [Wu, 2017], and the original RDF was defined [Hansen, 2013] as:

$$g_{AB}(r) = \frac{1}{\langle \rho_B \rangle_{\text{local}}} \times \frac{1}{N_A} \sum_{i \in A}^{N_A} \sum_{j \in B}^{N_B} \frac{\delta(r_{ij} - r)}{4\pi r^2} \tag{4.22}$$

$\langle \rho_B \rangle_{\text{local}}$ represents the particle of B density averaged over all shells around particle A.

Singh *et al.* studied the inhibition performance of three inhibitors AT-1, AT-2, and AT-3 on J55 steel in oil field formation water saturated with CO_2 [Singh, 2019]. They used the RDF technique to calculate the change in bond length and the inhibitor peak distance, considering a bond length between 1 and 3.5 Å representing chemical adsorption and a bond length greater than 3.5 Å representing physical adsorption, in reference to a previous work [Xie, 2015]. From showing the neutral and protonated inhibitors, the first peaks of all ATs-Fe were smaller than 3.5 Å, indicating chemical interactions, while two inhibitors in protonated form had physical interactions. The interactive forces were correlated with the binding energy and inhibition efficiency.

4.5.4 Mean-Square Displacement

Mean-square displacement (MSD) analysis is a computational method used to investigate the diffusion and mobility of particles in a system. MSD calculates the average displacement of particles over time, providing insights into their mobility and diffusivity. In the context of imidazoline inhibitors, MSD analysis can quantify the movement of inhibitor molecules near the metal surface and assess their ability to diffuse and explore the metal surface. The data can be used to interpret the results by evaluating the inhibitor's ability to migrate to reactive sites, estimating the diffusion coefficients, and understanding the influence of temperature or inhibitor concentration on molecular mobility. MSD analysis can give insights on the determination of inhibitor diffusion pathways, the identification of surface sites with high or low inhibitor mobility, and the characterization of inhibitor dynamics near defects or active corrosion sites. The limitations of MSD analysis are relevant to the need for sufficient simulation time to capture accurate diffusion behavior and the limitations imposed by the simulation time scale. Despite these limitations, MSD analysis provides valuable information on the mobility and diffusion characteristics of imidazoline inhibitors, contributing to the understanding of their corrosion inhibition mechanisms at the molecular level.

MSD analysis evaluates with the determination of the diffusion coefficients (D) the ability of inhibitors in restricting the migration of corrosive species to the metal surfaces. The lower the diffusion coefficients the higher the ability of the adsorbed inhibitor films in protecting the surface. The corrosive species in CO_2-saturated solutions commonly include HCO_3^-, CO_3^{2-}, Cl^-, H_3O^+, etc. The diffusion coefficient relations are mathematically given by [Babarao, 2008, Xie, 2015] as:

$$\mathrm{MSD}(t) = \left[\frac{1}{N} \sum_{i=1}^{N} \left| R_i(t) - R_i(0) \right|^2 \right] \tag{4.23}$$

$$D(t) = \frac{1}{6} \lim_{t \to \infty} \frac{\mathrm{d}}{\mathrm{d}x} \sum_{i}^{N} \left[\left| R_i(t) - R_i(0) \right|^2 \right] \tag{4.24}$$

$$D = \frac{m}{6} \tag{4.25}$$

$|R_i(t) - R_i(0)|^2$ is the mean-square displacement (MSD) and N is the number of target molecules, and the positions of the corrosive species at time 0 and t are given by $R_i(0)$ and $R_i(t)$. m is the MSD curve slope. Singh *et al.* evaluated the diffusion of CO_3^{2-}, Cl^-, and H_3O^+ through the adsorbed AT-1 layer [Singh, 2019]. The diffusion coefficients were respectively $0.043 \times 10^9\ m^2/s$, $0.023 \times 10^9\ m^2/s$, and $0.028 \times 10^9\ m^2/s$, which were higher than those through water reported from previous studies, suggesting that AT-1 acted as effective hindrance to the migration of the corrosive species studied, citing that methoxy groups constrain diffusion.

5

Physicochemical and Flow Effects

5.1 Effect of Chloride

The effect of chloride content on corrosion in oil pipelines is a critical factor to consider when studying the performance of imidazoline inhibitors. Chloride ions are highly corrosive and can initiate aggressive electrochemical reactions, leading to accelerated corrosion rates. As oil pipelines transport fluids that may contain chloride ions, understanding the impact of chloride content is crucial for effective corrosion protection. When chloride ions come into contact with the metal surface, they can promote the formation of corrosive species, such as ferric chloride. Ferric chloride can cause localized corrosion, particularly pitting corrosion, which can significantly compromise the integrity of the pipeline. Pitting corrosion occurs when small pits form on the metal surface, leading to the development of localized holes and potential leaks. Imidazoline inhibitors play a vital role in mitigating the corrosive effects of chloride ions. These inhibitors possess the ability to adsorb onto the metal surface and form a protective film. The protective film acts as a barrier, preventing chloride ions from reaching the metal and impeding the corrosion process. By effectively blocking the access of chloride ions to the metal surface, imidazoline inhibitors reduce the corrosion rate and enhance the longevity of the pipeline. The performance of imidazoline inhibitors in the presence of chloride ions depends on several factors. One crucial factor is the concentration of chloride ions in the environment. Higher chloride concentrations pose a greater risk of corrosion, requiring a more robust and effective inhibitor formulation. In addition to concentration, the chemical structure and adsorption affinity of imidazoline inhibitors are crucial considerations. The inhibitor's ability to adsorb onto the metal surface is essential for forming a protective film. Optimizing the inhibitor's formulation and chemical

Imidazoline Inhibitors for Corrosion Protection of Oil Pipeline Steels: Experimental Laboratory Evaluation and Case Studies, First Edition. Faysal Fayez Eliyan.
© 2024 John Wiley & Sons, Inc. Published 2024 by John Wiley & Sons, Inc.

structure can enhance its adsorption capabilities and improve its effectiveness in chloride-rich environments. Furthermore, the stability of the inhibitor film under chloride exposure is paramount. Chloride ions can destabilize the film, leading to film breakdown and increased corrosion rates. Therefore, it is crucial to select an inhibitor that can maintain its protective film integrity even in the presence of chloride ions. The evaluation of imidazoline inhibitors' performance in chloride-containing environments involves laboratory testing and field trials. Laboratory tests can simulate the corrosive conditions and measure the inhibitor's ability to reduce corrosion rates. Field trials are conducted in real-world pipeline environments to assess the long-term effectiveness of the inhibitor under actual operating conditions. To optimize the performance of imidazoline inhibitors in chloride-rich environments, a comprehensive understanding of the specific pipeline conditions is essential. This includes considering factors such as fluid composition, flow rate, temperature, and other corrosive species that may interact with chloride ions.

Chloride adsorbs physically on iron and steel surfaces, and generally occupies the free sites that are not covered with the imidazoline inhibitors. Its effect on imidazoline inhibiting the anodic reactions could be different from inhibiting the cathodic reactions – evaluated with potentiodynamic polarization of suitably selected scan rates within potential ranges that cover both the anodic and cathodic current densities. Zhang *et al.* investigated the effect of the chloride ions (introduced with 0.002, 0.1, and 2 mol/l) on the adsorption of amide imidazolines on iron with 0.2 mV/s potentiodynamic scans. They found that with the increase of chloride concentration the rate of the anodic reactions increased, the rate of the cathodic reduction decreased, and the inhibitor desorption potentials decreased, as shown in Figure 5.1.

5.2 Effect of Temperature

Temperature variations can significantly impact corrosion rates and the effectiveness of corrosion inhibitors in maintaining pipeline integrity. Understanding the relationship between temperature and corrosion is essential for designing effective corrosion protection strategies. Elevated temperatures can increase the rate of corrosion due to several factors. First, higher temperatures enhance the rates of chemical reactions, including oxidation and reduction reactions, which contribute to the corrosion process. Second, increased temperature results in greater mobility of corrosive species, facilitating their diffusion and reaction with the metal surface. Finally, thermal cycling, where the temperature fluctuates, can create thermal stress on the pipeline material, leading to cracking and accelerated corrosion.

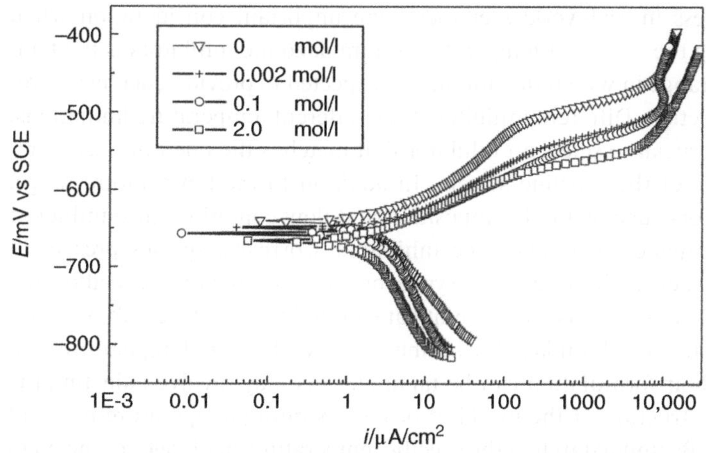

Figure 5.1 Potentiodynamic polarization of iron in 0.1 MPa-CO_2-saturated solutions containing 0.002, 0.1, and 2 mol/l chloride, in the presence of 20 ppm of amide imidazoline inhibitor and 0.5 $NaClO_4$ mol/l. *Source:* Zhang [2001]/with permission of Elsevier.

Imidazoline inhibitors are specifically designed to provide corrosion protection over a wide temperature range typically encountered in oil pipeline operations. These inhibitors function by adsorbing onto the metal surface and forming a protective film that acts as a barrier against corrosive agents. The protective film impedes the diffusion of corrosive species and reduces the corrosion rate. At elevated temperatures, imidazoline inhibitors must maintain their stability and effectiveness to provide reliable corrosion protection. The formulation of the inhibitors is crucial in determining their performance under varying temperature conditions. Factors such as inhibitor stability, adsorption kinetics, and film-forming properties are carefully considered during the design and optimization of imidazoline inhibitors. To evaluate the performance of imidazoline inhibitors under different temperature conditions, laboratory testing and field trials are conducted. Laboratory experiments simulate the corrosive conditions and measure the inhibitor's ability to reduce corrosion rates at various temperatures. These tests help in understanding the inhibitor's performance and effectiveness under controlled conditions. Field trials are conducted in actual pipeline environments to assess the long-term performance of the inhibitor under real-world operating conditions. These trials consider factors such as temperature fluctuations, thermal cycling, and the interaction of the inhibitor with other components in the pipeline system. Field trials provide valuable data on the inhibitor's performance over an extended period, helping to validate its effectiveness in real-world scenarios. One important consideration when evaluating the effects of temperature on imidazoline inhibitors is the temperature range in which the inhibitor is expected to provide effective corrosion protection. Different inhibitors have different temperature limitations, and it is crucial to select an inhibitor that matches the operating temperature range of the pipeline system. In addition to the temperature range, other factors such as fluid composition and flow conditions also influence the performance of imidazoline inhibitors. Corrosive species present in the fluid, such as chloride ions, oxygen, or acidic components, can affect the inhibitor's performance and stability at elevated temperatures. To optimize the performance of imidazoline inhibitors under different temperature conditions, their formulation can be tailored. This may involve adjusting the chemical structure of the inhibitor or incorporating temperature-resistant additives. By understanding the specific temperature challenges of the pipeline system and customizing the inhibitor formulation, it is possible to enhance the inhibitor's performance and ensure reliable corrosion protection across a wide temperature range.

For imidazoline inhibitors in general, increasing the temperature enhances its desorption from the surface, as investigated by Zhang *et al.*, indicated by the decrease in the desorption potentials of the potentiodynamic scans [Zhang, 2001]. In the same study, they found with electrochemical impedance spectroscopy (EIS) that the interface capacity showed an increase, suggesting that more surface areas were less covered by amide imidazoline at higher temperatures. Doležal *et al.* investigated the effect of temperature with the addition of different concentrations of imidazoline in CO_2-saturated solutions for a mild steel [Doležal, 2013]. Consistently, with the use of polarization and impedance methods, the inhibition efficiency decreased with higher temperature, regardless of the inhibitor concentration.

Jevremović *et al.* studied the inhibition performance of tall oil diethylenetriamine imidazoline (TOFA/DETA imidazoline) on the corrosion of a mild steel in CO_2-saturated in 3wt% NaCl solutions at different temperatures [Jevremović, 2016]. The addition of TOFA/DETA imidazoline shifted the corrosion potentials to more positive values, and as with higher temperatures the corrosion rates increased, but the inhibition efficiency did not interestingly change. The authors evaluated the influence of temperature with the Arrhenius equation, which relates the corrosion current density (i_{corr}) and temperature (T) to the activation energy of the reaction of transforming iron atoms into ions in the aqueous solution (E_a) as [Wang, 2015, Talbot, 1997]:

$$i_{corr} = A \exp\left(-\frac{E_a}{RT}\right) \tag{5.1}$$

The activation energy was determined from the slope of log i_{corr} versus $1/T$ plots. Increasing TOFA/DETA imidazoline concentration from 50 to 90 ppm decreased the activation energy from nearly 30 to 27 kJ/mol. The authors attributed that to a partial chemical adsorption of the inhibitor on the surface, similar to previous observations reported elsewhere [Bentiss, 2002, Amin, 2011, Deng, 2014]. The partial chemical adsorption was explained in terms of the inhibitor occupying sites of higher activation energy while sites of lower activation energy were left uncovered, leading to a collective lower activation energy.

Zhang *et al.* studied for API-X65 steel the efficiency of an imidazoline derivative in CO_2-saturated brines at different temperatures with potentiodynamic polarization [Zhang, 2007]. At test temperatures of 25, 45, 60, and 75 °C the corrosion rates of the inhibited surfaces were less than the uninhibited surfaces, indicating the efficiency of the inhibitor. The increase in the corrosion rates with higher temperature was a result of both the increase of the mass transfer processes and the reactive rate coefficient. The authors used equation 5.1 to calculate the activation energy in the absence (E_a) and presence (E_a^o) of the inhibitor, from equation 5.2.

$$R\ln\left(i_{corr}\right) = R\ln\left(k\right) - \frac{1}{T}E_a \qquad (5.2)$$

The relationship between ($R\ln(i_{corr})$) and ($1/T$) in the presence and absence of the inhibitor was plotted, producing liner relationships shown in Figure 5.2. Attained from the slopes of the lines, the activation energy in the absence

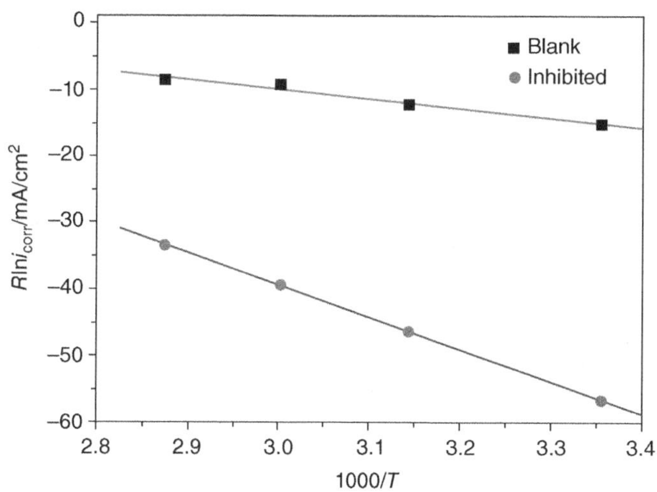

Figure 5.2 The relationship between ($R\ln(i_{corr})$) and ($1/T$) in the absence and presence of the inhibitor. *Source:* Zhang [2007]/with permission of Elsevier.

and presence of the inhibitor was respectively nearly 14 and 48 kJ/mol. The inhibitor affected the kinetics of both the anodic and cathodic reactions and raised the activation energy for the corrosion reactions.

Desimone *et al.* studied the corrosion inhibition mechanism of an amphiphilic amido-amine; *N*-[2-[(2-aminoethyl)amino]ethyl]-9-octadecenamide (AAOA) on mild steel surface in CO_2-saturated 5% NaCl solutions [Desimone, 2011]. They studied the effects of AAOA concentrations on the kinetics with potentiodynamic scans at 25, 35, and 45 °C for 54.4 μmol/dm^3 AAOA concentration. The anodic reactions increased with temperature but the anodic slopes did not change either in the uninhibited or inhibited solutions. The authors tabulated the kinetic parameters along with AAOA concentrations, in Table 5.1, mentioning the cathodic shift of the corrosion

Table 5.1 Electrochemical parameters calculated from polarization measurements on mild steel electrode in CO_2 saturated solution with various inhibitor concentrations at 25, 35, and 45 °C.

C (μmol/dm^3)	Temperature (°C)	j_{corr} (μA/cm^2)	C_R (μm/year)	η
0	25	65	750	—
	35	76	880	—
	45	105	1215	—
1.4	25	38	440	41
	35	52	610	31
	45	87	1010	17
2.7	25	25	290	61
	35	35	400	54
	45	50	580	52
8.2	25	6.3	75	90
	35	7.8	90	90
	45	9.3	110	91
13.6	25	1.3	16	98
	35	1.7	20	98
	45	1.6	18	99
27.2	25	0.79	9	99
	35	1.1	13	99
	45	1.6	18	99
54.4	25	0.45	5	99
	35	0.91	12	99
	45	1.4	16	99

Source: Adapted from Desimone [2011].

potentials with higher temperature. The anodic shift in the corrosion potentials with higher inhibitor concentrations indicated the preferential blocking of anodic reactions sites.

Depending on a previous study [Desimone, 2011], the significant change in the Tafel slope at concentrations higher than 8.2 μmol/dm^3 at 25 °C was attributed to micellation above critical micelle concentration (CMC) at which the inhibitor molecules in the solution would form a surfactant bi-layer (admicelle) on mild steel surface [Jovancicevic, 1998]. The authors highlighted the CMC at which the inhibitor changes the initial molecular solvated state to a state of different physical and chemical properties, with weak dependence on temperature. In order to obtain the CMC of AAOA, the surface tension (c) of various concentrations of the inhibitor was measured at 25 °C of CMC of AAOA of 15 μmol/dm^3 approximately. Above that concentration, the inhibitor is dispersible and forms stable emulsion-like layer; below it, the inhibitor is soluble and less effective.

The authors utilized the Arrhenius equation in reference to a previous study [Bentiss, 2005] to calculate the activation parameters from the corrosion current densities obtained from the polarization measurements. The activation energy of corrosion was obtained from Arrhenius plots at different inhibitor concentrations of 0.7, 1.4, 2.7, 8.2, 13.6, 27.2, and 54.4 μmol/dm^3. In the absence of the inhibitor the activation energy was nearly 33 kJ/mol, and between 0.7 and 2.7 μmol/dm^3 it was around 43 kJ/mol – ascribed to a formation of physically-adsorbed inhibitor [Conde, 2002, Vračar, 2002, Popova, 2003]. Activation energy increased to a maximum value at a concentration of 2.7 μmol/dm^3 and then decreased, attributed to a change in the adsorption behavior from being by physisorption at low concentrations. The three nitrogen atoms in the AAOA inhibitor molecule can accept a proton, exist in cationic form, and then adsorb on the negatively-charged metal surface, orienting horizontally or perpendicularly where the double-bond hydrocarbon chains can also interact with the surface through the π electrons.

At higher concentrations, the anodic sites were postulated to be blocked by chemisorption of the electron-rich amine or carbonilic moieties as the activation energy decreased [Bentiss, 2005, Okafor, 2009]. The authors suggested that mixed adsorption could take place as the molecules form a porous monolayer of lateral interactions. As the inhibitor concentration is higher, the surface is saturated with a protective bi-layer, consisting of interaction between iron of low-energy electron orbital and loosely bound electron with the nitrogen and oxygen heteroatoms in the inhibitor molecule, in a mechanism that was previously suggested [de Souza, 2009].

Wang *et al.* studied the corrosion behavior of J55 steel in high-pressure CO_2-saturated autoclave systems, at temperatures from 30 to 65 °C, in the presence of 80–150 ppm imidazoline inhibitor [Wang, 2014]. The corrosion behavior, along with the changes in the solution equilibrium chemistry, were studied for 72 hours in dynamic conditions of 0.3 m/s rotational speed in liquid phase and gas phase. They found that the corrosion rate increased with temperature to a maximum at nearly 60 °C, due to the formation of a protective corrosion film, and efficiently decreased with an inhibitor concentration greater than 150 ppm, as shown in Figure 5.3. Corrosion in

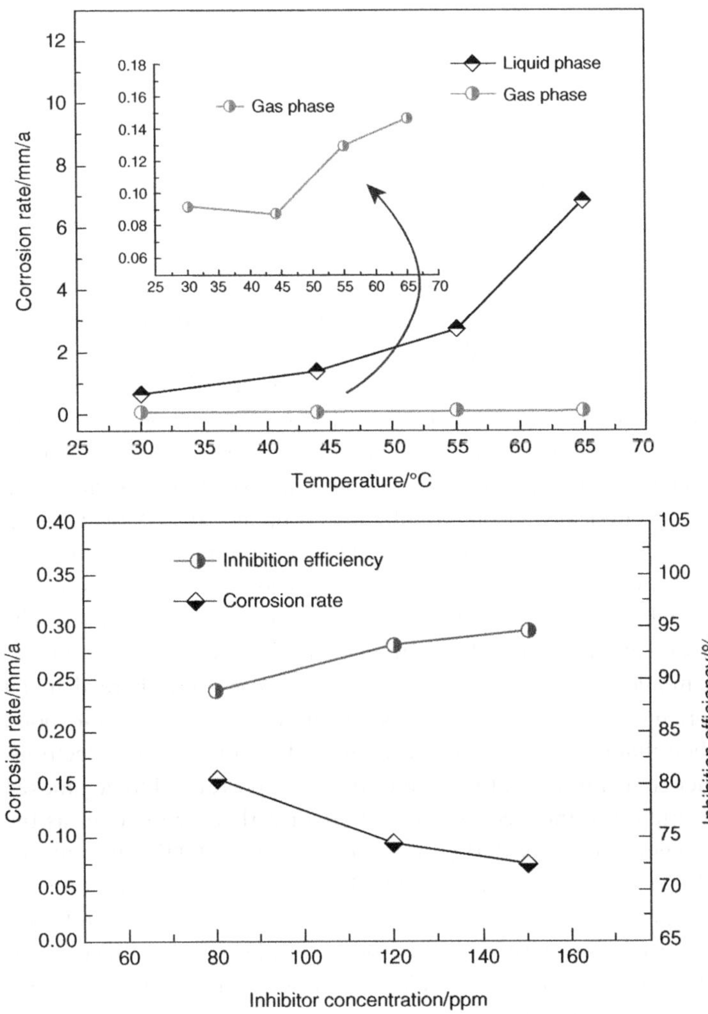

Figure 5.3 Influence of temperature on corrosion rates of J55 steel under simulated CO_2 flooding environment and influence of inhibitor on corrosion rate and inhibition efficiency of J55 steel under simulated CO_2 flooding environment. *Source:* Wang [2014]/Taylor & Francis.

gas phase was much lower than that in liquid phase. In both phases, temperature accelerated the kinetics and mass transfer of dissolution but also accelerated the formation and precipitation of protective iron-carbonate based products, decelerating corrosion rates with time [Videm, 1989]. The authors suggested that in the low-pH solutions an imidazoline-based inhibitor forms a layer that changes the oxidation reduction potential of hydrogen ions and complexes the oxidants of the surface. This was attributed to the NH_2 group that favors the formation of coordination bonds at the metal/inhibitor interface [Raichev, 1993, Cruz, 2004]. The authors studied the morphology of the corrosion products in relation to temperature and inhibitor concentration, outlined in Section 4.2 (Microscopic Evaluation).

Okafor *et al.* studied the effects of 25, 40, and 50 °C on the corrosion inhibition performance of 2-undecyl-1-ethylamino imidazoline (2UEI) on N80 mild steel in CO_2-saturated 3% NaCl solutions [Okafor, 2009]. The anodic current densities increased with temperature in the presence of 80 mg/l 2UEI, attributed to conventional reasons of the acceleration of the electrochemical, chemical, and transport processes at the interface. The authors reported an increase in the inhibition efficiency with temperature, attributing that to the enhanced chemisorption of the organic parts of the inhibitor molecules.

Ding *et al.* with the linear polarization resistance studied the influence of high temperatures of 120 and 150 °C on CO_2 corrosion rates of API 5L X65 carbon steel, with time in the absence and presence of DETA/TOFA imidazoline. At 120 °C, the corrosion rates decreased sharply with the injection of 440 ppm of the inhibitor to nearly stable 3 mm/year, compared to 10 mm/year decreasing gradually over eight hours of immersion to 1 mm/year [Ding, 2018]. At 150 °C, however, the corrosion rates were comparable to those of the uninhibited solution and inhibited solutions containing 440 and 880 ppm, with no mention of reasons by authors behind these trends.

In the double-autoclave system, in which the test samples and solutions were heated separately to isolate and investigate the effects of the inhibitor, the authors reported that the inhibitor at 120 °C, had the same effect, and at 150 °C, the inhibitor had no effect on the corrosion rate.

The authors analyzed the amounts of the inhibitors at the end of the tests at 120 and 150 °C, mentioning that they were similar in single- and double-autoclave systems [Ding, 2018].

Okafor *et al.* studied the effect of temperature of 20, 40, and 60 °C, on the corrosion inhibition for N80 steel, by rosin amide imidazoline (RAIM), in CO_2-saturated simulation formation solutions [Okafor, 2011]. In the presence of the 100-ppm inhibitor, the anodic and cathodic currents increased with temperature. The increase of inhibition efficiency with temperature indicated for the authors that the inhibition mechanism is chemical adsorption, otherwise it would be physical adsorption if the inhibition efficiency decreased. The authors calculated the activation energy value, which was slightly higher than the aqueous diffusion coefficient, indicating mixed interfacial reaction/diffusion control, and the decrease in activation energy with the addition of the inhibitor corroborated the finding that the adsorption mechanism is chemical. The heat of adsorption (Q_{ads}) was calculated as:

$$Q_{ads} = 2.303R\left[\log\left(\frac{\theta_2}{1-\theta_2}\right) - \log\left(\frac{\theta_1}{1-\theta_1}\right)\right] \times \left(\frac{T_1 \times T_2}{T_2 - T_1}\right) \tag{5.3}$$

where Q_{ads} is the heat of adsorption, R is the gas constant, and θ_1 and θ_2 are the degrees of surface coverage at T_1 (293 K) and T_2 (333 K), respectively, were determined using the equation:

$$\theta = \frac{i_{corr}^{o} - i_{corr}}{i_{corr}^{o}} \tag{5.4}$$

where i_{corr}^{o} and i_{corr} are the uninhibited and inhibited corrosion current densities, respectively. The heat of adsorption was of positive value, indicating that the adsorption process was endothermic and chemical-based.

5.3 Flow Effects

Flow effects play a significant role in the corrosion behavior of oil pipelines and the performance of imidazoline inhibitors. Understanding how flow conditions impact corrosion rates is crucial for designing effective corrosion protection strategies in pipeline systems. Flow in oil pipelines can vary in terms of flow rates, flow regimes, and the contents of the flowing fluid. These factors collectively influence the corrosion process and the performance of imidazoline inhibitors. Flow rates have a direct impact on corrosion rates. Higher flow rates can result in increased turbulence, leading to more aggressive corrosion. Turbulent flow causes localized fluid motion and creates eddies, which can remove protective films from the metal surface and expose it to corrosive agents. Imidazoline inhibitors are designed to mitigate the adverse effects of flow on corrosion. When applied in pipelines with turbulent flow, these inhibitors adsorb onto the metal surface and form a uniform and robust film. The protective film acts as a shield, protecting the metal from the detrimental effects of turbulence and flow-induced corrosion.

Flow regimes also play a crucial role in corrosion. Flow can be classified into laminar or turbulent regimes, depending on the Reynolds number, and the content and velocity of different phases. In laminar flow, where fluid motion is smooth and predictable, imidazoline inhibitors excel at forming a stable and continuous film on the metal surface. This film acts as a barrier, preventing the diffusion of corrosive species and reducing corrosion rates. By adhering to the metal surface, the inhibitors provide long-lasting corrosion protection even in the absence of turbulent fluid motion. Turbulent flow poses a significant challenge in corrosion protection due to increased shear forces and fluid velocities. Imidazoline inhibitors have been developed to withstand these harsh conditions by adsorbing onto the metal surface and forming a robust and adherent film. This protective film shields the metal from the disruptive effects of turbulence, maintaining corrosion resistance and preventing localized corrosion. In the presence of stratified flow, where distinct liquid and gas layers exist, imidazoline inhibitors ensure corrosion protection by targeting the liquid phase. The inhibitors form a protective film on the metal surface exposed to the liquid phase, reducing corrosion rates and minimizing the risk of localized corrosion. By addressing the corrosive nature of the liquid phase, imidazoline inhibitors contribute to overall pipeline integrity. Slug flow, characterized by intermittent gas slugs within the liquid phase, can result in fluctuating flow patterns and increased corrosion rates. Imidazoline inhibitors are designed to mitigate the effects of slug flow by forming a protective barrier on the metal surface. The inhibitors create a stable film that reduces the corrosive action of the liquid slug regions and the erosive effects of the gas slugs, providing effective corrosion protection. Annular flow, with a continuous liquid film along the pipe wall and gas flowing in the core region, can present varying corrosion challenges. Imidazoline inhibitors help maintain a stable and uniform liquid film, which acts as a protective barrier against corrosion. By reducing the impact of variations in the liquid film thickness and stability, the inhibitors enhance the corrosion resistance of the pipeline. Mist flow, characterized by dispersed liquid droplets within the gas phase, can result in localized corrosion when the droplets collide with the pipe wall. Imidazoline inhibitors combat the corrosive effects of mist flow by forming a protective film on the metal surface. This film reduces the corrosive action of the dispersed liquid droplets and helps prevent localized corrosion. Despite the presence of sand in some pipeline flows, imidazoline inhibitors have the ability to adsorb effectively onto the metal surface, ensuring continuous protection against corrosion. Additionally, when combined with dispersants, these inhibitors work synergistically to prevent sand particle agglomeration and deposition, further improving the overall performance in sand-containing flows. Compatibility with sand control measures, such as sand screens or gravel packs, is crucial, and imidazoline inhibitors are designed to be compatible with these measures, ensuring corrosion protection for exposed metal surfaces. The performance of imidazoline inhibitors in sand-laden flows is evaluated through laboratory testing and field trials, where the inhibitors' corrosion inhibition efficiency is assessed under simulated flow conditions and real-world operating conditions, respectively.

Jiang *et al.* through EIS experiments evaluated the effects of the flow rates on the polarization resistance, from which the mass loss of N80 steel was calculated in the presence of 100 mg/l imidazoline inhibitor [Jiang, 2005A]. Liu *et al.* evaluated the performance of three inhibitors, thioglycolic acid (TGA), diethylenetriamine (DETA), and naphthene acid imidazoline (IM) on bare surface and pre-corroded N80 steel in CO_2-saturated 1%NaCl solutions, at 35 °C [Liu, 2011]. The test samples were rotating electrodes in an autoclave and the performance was evaluated with weight loss and electrochemical methods. In polarization experiments, in Figure 5.4, the bare surface showed a decrease in the anodic and cathodic currents with potentials of positive shift, highest by IM inhibitor both in stagnant and rotation conditions. The inhibition efficiency of IM inhibitor did not significantly change with the rotation, as outlined in Table 5.2, and the corrosion resistance for the pre-corroded surfaces decreased further with the rotation speed in the presence of IM inhibitor.

Ortega-Toledo *et al.* examined the effect of flow with a number of electrochemical techniques in the presence of 8.1×10^{-5} mol/l carboxyethyl-imidazoline (recommended from field observations) in 3% NaCl CO_2-saturated solutions at 50 °C [Ortega-Toledo, 2010]. The material, introduced as cylindrical specimens, was a new HSLA API X-120 pipeline steel of the elements shown in the study. The corrosion rates increased with rotational speeds of 250, 500, 1000, and 2500 rpm, and the potentiodynamic scans showed that both the corrosion potentials and corrosion current densities increased with rotation speed. The authors attributed the abrupt changes in the anodic branches, compared with those of the uninhibited conditions, to the anodic desorption of the inhibitor at specific potentials with rotation speeds [Hong, 2002]. From the kinetic data of the uninhibited conditions (followed Levich equation) and inhibited conditions (did not follow Levich equation); the authors suggested that the corrosion reactions were under diffusion control and partial diffusion control, respectively. Up to 500 rpm, the inhibitor desorbed at as high as −530 mV vs. SCE; beyond that rotational speed the inhibitor was unstable. The authors tabulated the limiting current densities in the anodic and cathodic branches and monitored the variation in the polarization resistance with time at different speeds in uninhibited and inhibited solutions. The authors indicated the basic mechanism of covering the surface and forming metal-inhibitor complexes that protect the surface from impedance spectra. They cited that at high

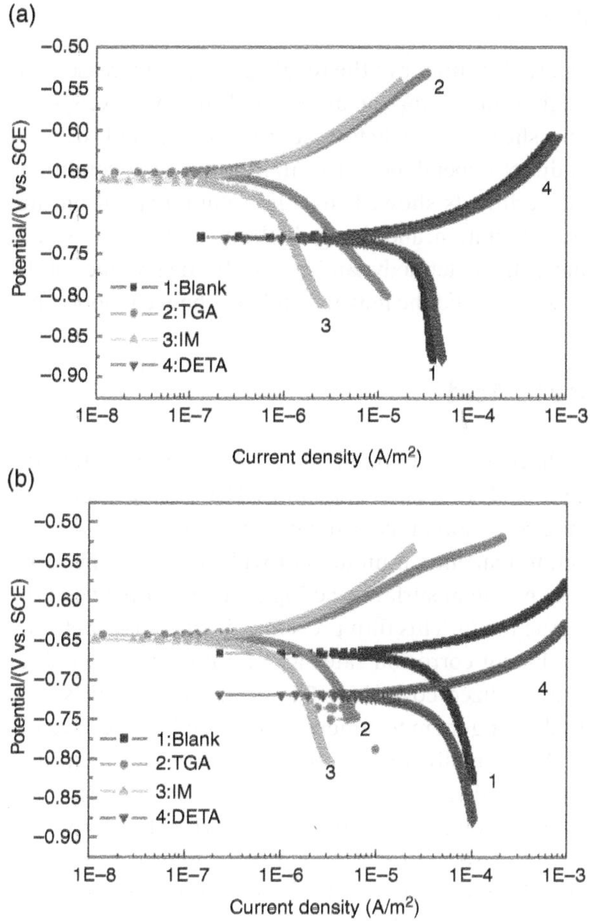

Figure 5.4 Polarization curves of N80 steel in CO_2-saturated 1% NaCl solution with 50 mg/l inhibitors at 35 °C at (a) 0 rpm and (b) 2000 rpm. *Source:* Liu [2011]/ John Wiley & Sons.

Table 5.2 Effect of rotating speed on the inhibition efficiency for bare surface of N80 steel in CO_2-saturated 1% NaCl solution with 50 mg/l inhibitors at 35 °C.

	Corrosion rates (mm/year)		Inhibition efficiency (%)	
Inhibitors	**0 (rpm)**	**2000 (rpm)**	**0 (rpm)**	**2000 (rpm)**
Blank	2.694	4.358	—	—
TGA	0.105	0.178	96.1	95.9
IM	0.053	0.089	98.0	97.9
DETA	2.834	4.318	−5.2	0.92

Source: Liu [2011]/John Wiley & Sons.

flow conditions, transport of inhibitor to the reacting surface increases, but it could be adversely high in increasing the diffusion of the corrosive species and in removing by flow shear the newly formed protective complexes, and this explained the no direct dependence of inhibition on flow conditions [Branzoi, 2000]. EIS experiments showed that the rotation speed did not change the mechanism of inhibition, and change in the polarization resistance was consistent with that of the potentiodynamic scans. In the presence of the inhibitor, the mass loss was less and the flow rate did not change it noticeably.

5.4 Effect of Acetic Acid

Imidazoline inhibitors have proven to be highly effective in mitigating the corrosive effects of acetic acid (HAc) in oil pipelines. HAc, being a corrosive agent, can accelerate the corrosion process and compromise the integrity of the pipelines. However, imidazoline inhibitors provide a solution by forming a protective film on the metal surfaces, acting as a barrier between the corrosive medium and the metal. This film prevents direct contact between HAc and the metal, reducing corrosion rates and minimizing the risk of metal loss, pitting, and localized corrosion. The inhibitors exhibit strong adsorption properties, allowing them to adhere to the metal surface even in the presence of HAc. Additionally, imidazoline inhibitors can neutralize the acid, further enhancing their corrosion inhibition capabilities. Their ability to provide long-term protection ensures the integrity of the pipelines over extended periods of exposure to HAc. Field performance and testing have confirmed the effectiveness of these inhibitors, demonstrating significant reductions in corrosion rates and improved pipeline integrity in real-world conditions. Compliance with regulatory standards and environmental regulations is also a crucial aspect, and imidazoline inhibitors meet these requirements, ensuring the highest level of safety and environmental protection. Li *et al.* reported the effects of 0.1 g/l of 1-(2-aminoethyl)-2-heptadecyl imidazoline, in the presence of 1 and 3 g/l HAc on the corrosion of N80 carbon steel in CO_2-saturated 1.65 wt% NaCl solutions with electrochemical methods [Li, 2017]. They reported that adding the inhibitor decreased both the anodic and cathodic currents with both HAc concentrations. In 1 g/l HAc solution, the inhibitor was predominantly anodic, and in 3 g/l HAc solution, the inhibitor was predominantly cathodic. The authors extracted the potentiodynamic parameters from the scan profiles, showing the significant effects of the inhibitor in both HAc solutions. Without addition of the inhibitor, the Nyquist plots were composed of a high-frequency capacitive loop (the double layer) and a low-frequency inductive loop (adsorbed intermediate products). In the inhibited solutions, the impedance was much higher than that of the uninhibited solution, and higher in the 3 g/l HAc solution than that of 1 g/l HAc solution, without change in the interactions mechanisms, and the EIS parameters were extracted and tabulated.

5.5 Effect of H_2S

Hydrogen sulfide (H_2S) is a highly corrosive gas commonly found in oil and gas production, transportation, and refining processes. It poses a significant challenge to the integrity of oil pipelines and necessitates effective corrosion control measures. Imidazoline inhibitors have been extensively studied and developed to address the corrosive effects of H_2S and provide reliable protection against H_2S-induced corrosion. H_2S is known to be a corrosive gas that can cause severe damage to metal surfaces. When present in oil pipelines, H_2S can react with moisture and form sulfuric acid, which attacks the metal, leading to sulfide stress corrosion cracking (SSC), hydrogen embrittlement, and general corrosion. In order to mitigate the damaging effects of H_2S, corrosion inhibitors, such as imidazoline inhibitors, are crucial. Imidazoline inhibitors play a vital role in inhibiting H_2S-induced corrosion. They function by forming a protective film on the metal surface, acting as a barrier against corrosive species. This film impedes the interaction between H_2S and the metal, effectively reducing corrosion rates and minimizing the risk of structural failures. The stability of the film is of utmost importance, as it ensures continuous corrosion inhibition, even under fluctuating conditions or in the presence of varying H_2S concentrations. In addition to film formation, imidazoline inhibitors can exhibit H_2S scavenging properties, further enhancing their corrosion inhibition capabilities. In some cases, imidazoline inhibitors are used in combination with other corrosion inhibitors to enhance the overall corrosion protection. Certain additives, such as mercaptobenzothiazole (MBT), can work synergistically with imidazoline inhibitors to provide improved H_2S corrosion inhibition. The combination of inhibitors can offer a broader range of protection and better performance under varying operating conditions. Extensive laboratory and field testing, along with compliance with regulatory standards, further validate their performance. Continuous monitoring and maintenance practices ensure long-term corrosion control and the integrity of oil pipelines operating in H_2S-containing environments. Javidi *et al.* investigated the effectiveness of 50 ppm amine-based imidazoline on corrosion inhibition of API 5L X65 carbon steel, in the presence of iron carbonate scale and H_2S corrosion products, in CO_2-saturated 3.5 wt% NaCl solutions [Javidi, 2019]. The material was cut from a wet gas transmission pipeline in service for 12 years. As shown in Figure 5.5 below, increasing the concentration of H_2S increased

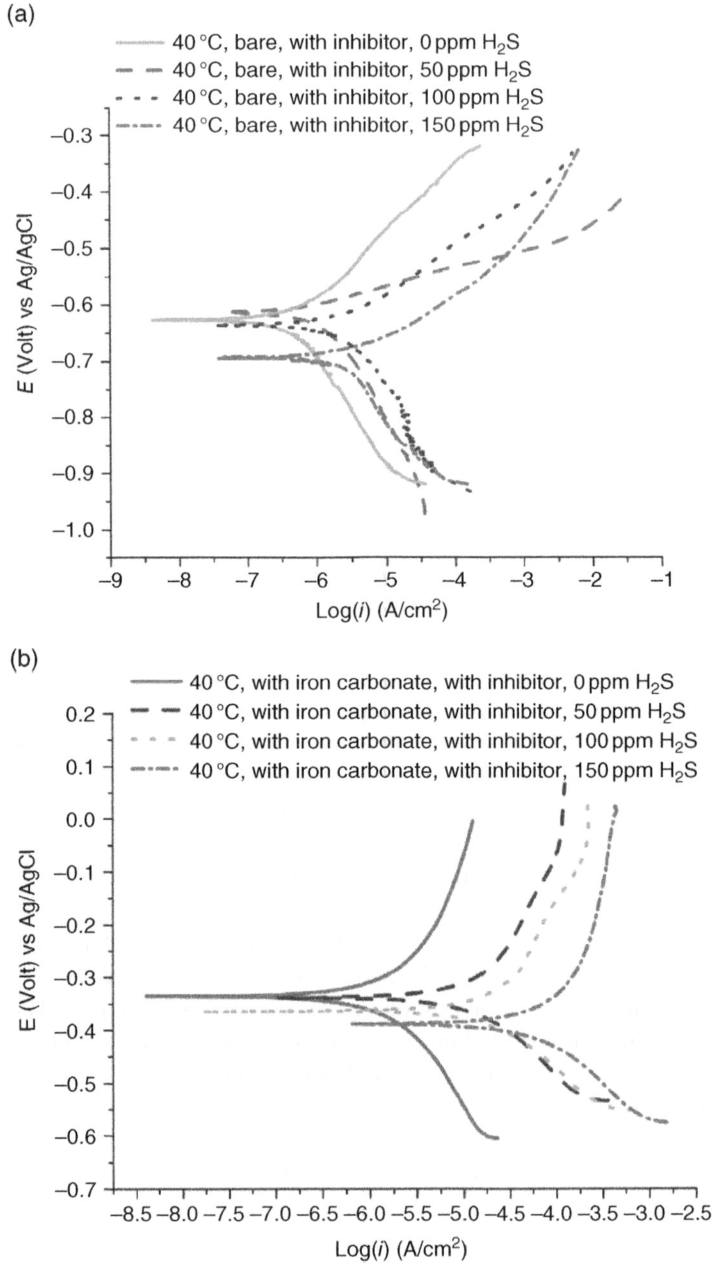

Figure 5.5 Tafel polarization curves for (a) Bare samples inhibitor; (b) Samples with iron carbonate scale in CO_2-saturated 3.5 wt% NaCl solution containing 0, 50, 100, 150 ppm H_2S, and 50 ppm inhibitor. *Source:* Adapted from Javidi [2019].

the corrosion rates and cathodic reduction in the presence and absence of the corrosion products. In the presence of the corrosion products the anodic reaction current densities were less and the corrosion potentials were higher. The inhibition efficiency ranged from 85% to 99%, and from 45% to 85%, respectively, for bare samples and samples with corrosion products, of inhibitor concentrations correspondingly from 150 to 50 ppm. In addition, from EIS measurements, the corrosion mechanism did not seem to be influenced by the concentration of H_2S, but increasing its concentration, confirming with the polarization results, reduced the charge-transfer resistance.

5.6 Metallurgical and Surface Effects

Metallurgy and surface characteristics play a vital role in the effectiveness of imidazoline inhibitors for corrosion prevention and control in pipeline steels. The choice of steel composition and the condition of the surface greatly influence the interactions between the inhibitor and the steel substrate, ultimately impacting the corrosion inhibition performance.

Pipeline steels used in the oil and gas industry are typically selected to possess specific metallurgical properties that ensure strength, toughness, and resistance to various environmental conditions. These steels are commonly classified as carbon steels, low-alloy steels, or stainless steels, each having distinct metallurgical characteristics. Carbon steels, widely used in pipelines, consist mainly of iron and carbon, with small amounts of other elements. The presence of carbon imparts strength to the steel, but it also renders it susceptible to corrosion. The surface of carbon steels usually contains iron oxide layers, mill scale, or other contaminants, which can hinder the interaction between the imidazoline inhibitor and the steel surface. Effective cleaning and surface preparation techniques are crucial to remove these layers and enhance inhibitor adsorption. Low-alloy steels, containing additional alloying elements such as manganese, chromium, nickel, or molybdenum, offer improved mechanical properties and corrosion resistance compared to carbon steels. The presence of alloying elements influences the microstructure of the steel, affecting its corrosion behavior. For instance, chromium and molybdenum enhance the steel's resistance to localized corrosion, while nickel improves its resistance to general corrosion. Imidazoline inhibitors can interact differently with these alloying elements, leading to variations in the inhibition efficiency. Stainless steels, known for their excellent corrosion resistance, contain a high percentage of chromium, which forms a protective passive oxide layer on the surface. The surface characteristics of stainless steels, such as surface finish and cleanliness, significantly impact inhibitor adsorption and corrosion inhibition. A smooth and clean surface promotes the formation of a protective film by the imidazoline inhibitor, effectively preventing corrosion. Surface preparation techniques, such as abrasive blasting, acid cleaning, or pickling, are commonly employed to remove scale, oxides, and contaminants from the steel surface. These techniques not only clean the surface but also create a roughness profile that enhances inhibitor adsorption. The roughness and cleanliness of the steel surface influence the inhibitor's ability to penetrate the oxide layer, establish a protective film, and inhibit corrosion effectively. Surface modifications, such as coatings or surface treatments, can further enhance the performance of imidazoline inhibitors. Coatings, such as epoxy or polymeric films, provide an additional barrier between the steel surface and the corrosive environment. Surface treatments, such as phosphating or passivation, modify the surface chemistry, improving the adsorption and film-forming capabilities of the inhibitor. Zhang *et al.* evaluated the inhibition performance of an imidazoline inhibitor, for two types of steels, X70 and Q345 steels, in CO_2-saturated 3 wt% brines [Zhang, 2017]. The chemical compositions of the steels are in Table 5.3, and their microstructures are shown in Figure 5.6. The alloying elements Mo, Cr, Nb, Ti, and Cu adds to the strength of the steel and its corrosion resistance. The microstructure of Q345 steel has higher carbide content than X70, as reflected from their

Table 5.3 Typical chemical compsotions of X70 and Q235 steels.

Element	C	Si	Mn	Mo	Cr	Nb	Ti	V	Ni	Cu	Fe
X70	0.064	0.18	1.5	0.17	0.019	0.025	0.016	0.026	0.12	0.11	Balance
Q235	0.19	0.25	0.5	—	—	—	—	—	—	—	Balance

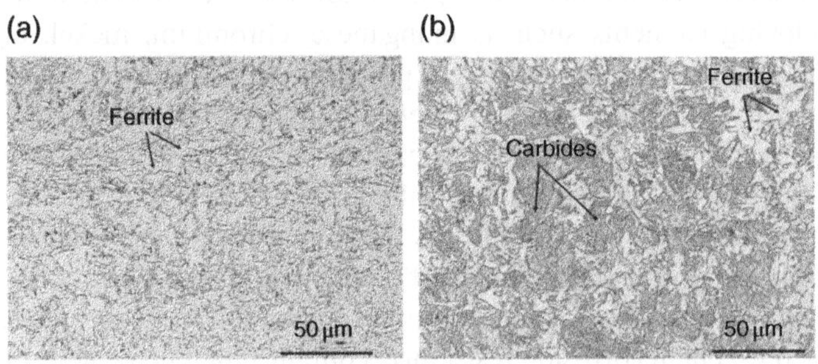

Figure 5.6 Microstructures of (a) X70 and (b) Q235 steels. *Source:* Zhang et al., [2017]/with permission from Elsevier.

microstructures. The metallurgical effects of the two test steels were manifested in both the weight loss and electrochemical measurements. The open-circuit potentials of two steels with different concentrations of the inhibitor were monitored for 3600 seconds at 333 K (60 °C).

The open-circuit potentials increased with the inhibitor concentration, which was of mixed-type, blocking the active sites of dissolution. The open-circuit potentials, regardless of temperature, indicated that the X70 was more active than Q235, and the inhibition efficiency for X70 was lower than Q235 due to its ferrite content. Separate weight loss measurements for 96 hours were conducted for the two steels in the same conditions to examine the surface morphologies without and with the addition of 4.8780×10^{-5} mol/l of the inhibitor. Both surfaces were covered with corrosion products, but with the addition of the inhibitor the surfaces were protected with an adsorbed inhibitor film that formed from an early period of immersion, achieving interestingly comparable inhibition efficiencies. Decreasing the inhibitor concentration to 4.8780×10^{-6} mol/l, however, resulted in different behavior. Q235 was well protected and X70 was severely damaged, with inhibition efficiencies of nearly 95% and 8%, respectively, suggesting that the inhibitor achieved better adsorption – in reference to the preferential adsorption onto the carbides (Fe_3C) of higher proportion in Q235, via Lewis acid interactions [Lopez, 2004]. Zhang *et al.* studied the inhibition performance of an imidazoline inhibitor on CO_2 corrosion of a carbon steel, in 3 wt% NaCl 60 °C solutions, of 0.18 wt% C, 0.25 wt% Si, 0.5 wt% Mn, of two microstructures [Zhang, 2018]. The first microstructure was produced by annealing, austenized at 850 °C and furnace cooled, having coarse laminar pearlite, denoted as H steel. The second microstructure was produced by quenching and tempering, austenized at 850 °C, water quenched and tempered for one hour at 630 °C, having coarse laminar pearlite, denoted as T steel. The authors considered a matrix of experimental conditions for weight loss measurements, with two different concentrations of the inhibitor and two different time periods of immersion, and the average corrosion rates (in mm/year) and inhibition efficiencies are shown in Figure 5.7. The corrosion rates were calculated for samples of removed corrosion products by immersion after the experiments in 1 l HCl + 20 g Sb_2O_3 + 50 g $SnCl_2$ (ISO 8407: 1991) for 10 minutes, according to a procedure mentioned in [Xu, 2016], with the utilization of the equation, used previously for similar inhibition investigation [Umoren, 2015]:

$$W = \frac{8.76 \times 10^4 \Delta m}{\rho} \tag{5.5}$$

W is the general corrosion rate of the steel (mm/year), S is the electrode area (cm^2), t is the immersion time, Δm is the weight loss (g), and q is the

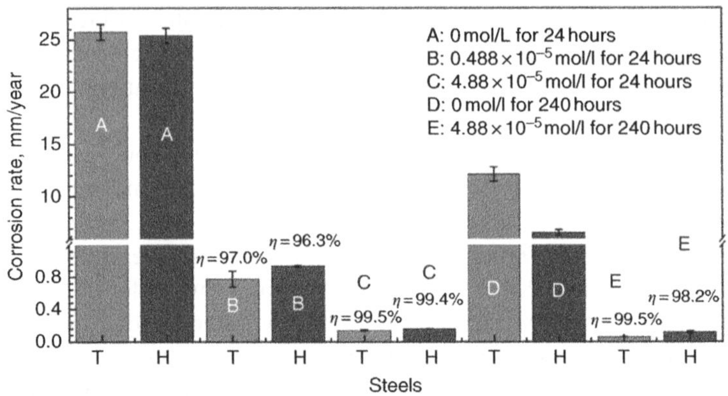

Figure 5.7 General corrosion rates (obtained from weight loss measurements) and inhibition efficiencies for samples after immersion in 3 wt% NaCl solution containing CO_2 with various concentrations of inhibitor at 60 °C. *Source:* Zhang [2018]/with permission of Elsevier.

steel density (g/cm^3). For localized corrosion, a 3D laser microscopy (LEXT OLS4100, OLYMPUS) was used to detect the depth of ten deepest pits as an average for use in the equation [NACE, 1999, Wei, 2016]:

$$V_{\text{corr}} = \frac{365 \times 24 \times h}{t} \tag{5.6}$$

V_{corr} is the localized corrosion rate (mm/year), h is the average corrosion depth (mm), and t is the immersion time (hours). The localized corrosion rates were calculated for selected conditions. In the uninhibited solutions, the corrosion rates were comparable, of nearly 25 mm/year, between the two steels after 24 hours of immersion, but after 240 hours, the corrosion rate of H steel was nearly twice as high as that of T steel. The difference in the corrosion rates was ascribed to the difference in the exposed areas of Fe_3C, as traditionally the trend would be ascribed to [Ueda, 1996]. Fe_3C in laminar structure (T steel) has better anchoring effect for the corrosion products than the granular structure (H steel), and causes higher corrosion rates. In the presence of the inhibitor during 24 hours, the corrosion rates were much less for both steels with high efficiency with 0.488×10^{-5} mol/l and 4.88×10^{-5} mol/l concentration. For 240 hours, the corrosion rates were increasingly less, and it seemed that the inhibitor upon adsorption was more protective on T steel than on H steel in the long term. The localized corrosion rates for H steel were higher than that of T steel, and they were more sensitive to the inhibitor concentration.

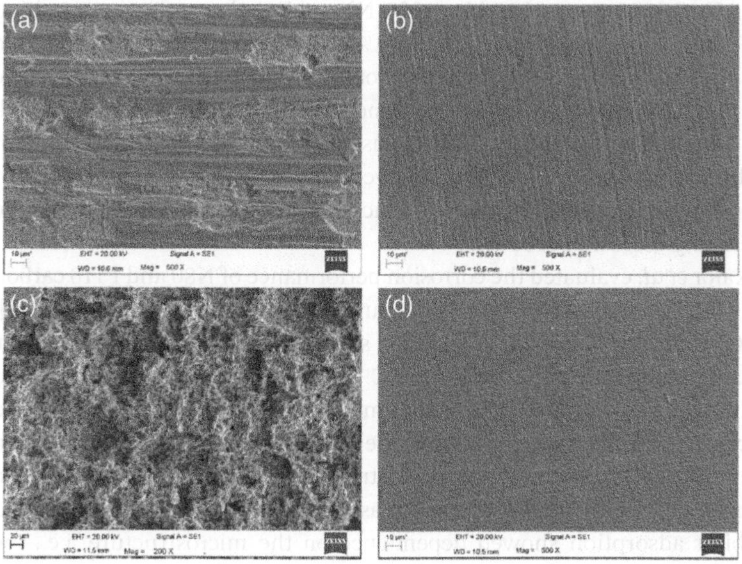

Figure 5.8 SEM images of samples (with different initial surface roughness) after removing the scale in the presence of 1.22×10^{-5} mol/l inhibitor for different time: (a) 150#, 1 hour; (b) 5000#, 1 hour; (c) 150#, 24 hour; (d) 5000#, 24 hour. *Source:* Zhang et al., [2019]/with permission of Elsevier.

Zhang *et al.* examined the effects of surface roughness of 150 and 5000 grit of X70 steel in the presence of 1.22×10^{-5} (in Figure 5.8) and 1.22×10^{-4} mol/l of thioureido imidazoline with different immersion times of 1, 24, and 240 hours [Zhang, 2019]. They found the damage was more severe on rougher surfaces, regardless of the immersion time. For longer periods of immersion for 240 hours, the authors made the inhibitor concentration 10 times greater, and the surface damage was less, but the corrosion attack was localized and pits were observed, attributed to the roughness not to inclusions. The authors proposed that the inhibitor molecules would preferentially adsorb better onto smoother surfaces making more stable inhibitor films, especially in low-concentration solutions.

Wang *et al.* used weight loss and electrochemical techniques to study the significance of galvanic coupling on the corrosion behavior of, and inhibition performance of three inhibitors on N80 steel and 13Cr stainless steel [Wang, 2019]. The three inhibitors were imidazoline (IM), 1-butyl-2-thiobe nzylbenzimidazole (BTBBI) and lauric acid (LA), with concentrations of 100 mg/l, in dynamic supercritical CO_2 oilfield water conditions. The composition (wt%) of N80 CS was: 0.34% C, 0.2% Si, 1.45% Mn, 0.02% P, 0.015% S,

0.15% Cr, 0.008% Cu, 0.018% Mo, 0.03% Ni, and Fe balance. The composition (wt%) of 13Cr SS was: 0.08% C, 1% Si, 1% Mn, 0.035% P, 0.03% S, 13.05% Cr, 0.5% Ni, and Fe balance. The Nyquist plots for both materials were capacitive and the significance of surface inhomogeneity and dispersion appeared. Two different equivalent circuits were used, however, to fit the data, without attributing the capacitive and inductive loops to specific characteristics of the inhibitors, or to the nature of interactions between the inhibitors, or the developing corrosion products.

Okafor *et al.* evaluated the corrosion performance of N80 and P110 carbon steels, with compositions and AFM analysis outlined in [Okafor, 2011], in CO_2-saturated simulation formation solutions in the presence of 50, 100, and 200 ppm RAIM, at 20, 40, and 60 °C. The SEM micrographs showed that the inhibitor performed well for both metals, with the absence of localized attacks or rough surfaces induced by the corrosive agents in test solutions in different ways according to the microstructure. The EIS analysis, however, revealed that inductive behavior that associates to the mass transport and inhibitor adsorption showed dependence on the microstructure, i.e., the corrosion mechanisms for both carbon steels were different, or the inhibitor influenced the same corrosion mechanism in different ways. Also, the potentiodynamic data of the corrosion current densities and Tafel parameters showed appreciable discrepancies among the two carbon steels at different inhibitor concentrations.

6

Case Studies: Laboratory Evaluations

6.1 Experimental Evaluation of Tetraethylenepentamine (TEPA) Imidazoline Inhibitor for Effective Corrosion Control in an Oil Pipeline: Electrochemical and Spectroscopic Methods

The case study focuses on an oil pipeline company that has undertaken a comprehensive program to evaluate the effectiveness of Tetraethylenepentamine (TEPA) imidazoline inhibitor for corrosion control. The company aims to ensure the long-term reliability and integrity of the oil pipeline by implementing experimental evaluation methods that include both electrochemical and spectroscopic techniques.

To assess the corrosion control efficiency of the TEPA Imidazoline inhibitor, various electrochemical methods were employed. These methods included electrochemical impedance spectroscopy (EIS), which provided valuable information about the impedance, corrosion rate, and charge transfer resistance. Potentiodynamic polarization scans were conducted to determine the corrosion potential and corrosion current density, allowing for a comprehensive understanding of the corrosion behavior. Additionally, cyclic voltammetry experiments were performed to identify the inhibitor's redox behavior and its impact on the corrosion process. The electrochemical methods were conducted with specific parameters to ensure realistic evaluation. The frequency range for EIS analysis was set from 0.01 Hz to 100 kHz, with an amplitude of 10 mV. Potentiodynamic polarization scans were performed at a scan rate of 1 mV/s, enabling accurate measurement of the corrosion parameters. Cyclic voltammetry experiments were

Imidazoline Inhibitors for Corrosion Protection of Oil Pipeline Steels: Experimental Laboratory Evaluation and Case Studies, First Edition. Faysal Fayez Eliyan.
© 2024 John Wiley & Sons, Inc. Published 2024 by John Wiley & Sons, Inc.

conducted at a sweep rate of 50 mV/s, facilitating the identification of the inhibitor's electrochemical behavior. Through cyclic voltammetry, the voltammetric response of the TEPA Imidazoline inhibitor could be analyzed, providing insights into its electrochemical properties. The technique enabled the determination of key electrochemical parameters, such as peak potentials, peak currents, and scan rates, which provided valuable information about the inhibitor's redox behavior. The obtained cyclic voltammograms allowed researchers to evaluate the inhibitor's ability to undergo oxidation and reduction processes, shedding light on its stability, reactivity, and overall performance as a corrosion inhibitor. The technique also facilitated the detection of any potential side reactions or competing reactions that could influence the inhibitor's effectiveness. Additionally, cyclic voltammetry provided a means to assess the kinetics of the electrochemical reactions associated with the TEPA Imidazoline inhibitor. By analyzing the shape, position, and intensity of the peaks in the cyclic voltammograms, information regarding the reaction rates and mechanisms could be inferred, aiding in the understanding of the inhibitor's behavior under different conditions.

In addition to electrochemical methods, spectroscopic techniques were employed to evaluate the performance of the TEPA Imidazoline inhibitor. Fourier-transform infrared spectroscopy (FTIR) was used to analyze the inhibitor's chemical bonds and functional groups in the mid-infrared region. Specific absorption peaks in the FTIR spectrum corresponded to vibrational modes of different functional groups, such as N—H, C—H, C=O, and C—N. The positions, intensities, and shapes of these peaks provided valuable insights into the inhibitor's molecular structure and confirmed the presence of specific chemical moieties. Furthermore, FTIR spectroscopy enabled researchers to analyze the interactions between the TEPA Imidazoline inhibitor and the steel surface. By comparing FTIR spectra before and after the inhibitor's exposure to the steel surface, any shifts or changes in the absorption bands could be observed, indicating the formation of a protective film or the interaction between the inhibitor and the steel. This allowed for the identification of key molecular characteristics and interactions with the steel surface. X-ray photoelectron spectroscopy (XPS) was employed to determine the elemental composition and chemical states on the steel surface, providing insights into the inhibitor's effectiveness in preventing corrosion. By subjecting the steel surface to X-ray radiation, XPS provided information about the binding energies of the emitted photoelectrons, which directly correlated to the elemental composition of the surface. The obtained XPS spectra enabled the identification and quantification of different elements present, including iron (Fe), oxygen (O), carbon (C), nitrogen (N), and any additional elements that might be present due to surface contamination.

The laboratory experiments were conducted under realistic environmental conditions to simulate the oil pipeline's operational environment. The temperature was set to match the average temperature experienced by the pipeline, typically around 50 °C. The pH of the solution used in the evaluation was adjusted to resemble the pH of the transported oil, which typically ranged from 6.5 to 7.5. This pH range was selected to accurately represent the corrosive conditions experienced within the oil pipeline. Furthermore, the presence of water, H_2S, and CO_2 was considered during the evaluation. The water phase in the transported oil had specific properties, including a pH of 7.0, salinity levels of 10,000 ppm, and dissolved oxygen content of 2 mg/l. The concentration of H_2S in the solution was set at 5 ppm, representing the typical levels encountered in the oil pipeline. Similarly, the concentration of CO_2 was maintained at 10,000 ppm to simulate the corrosive effects of CO_2 present in the transported oil. The oil being transported through the pipeline was classified as light crude oil and possessed specific properties that included a specific density ranging from 0.78 to 0.85 g/cm^3, viscosity within the range of 10–25 cSt, sulfur content of approximately 0.1–0.5 wt%, and acidity levels between 0.5 and 1.0 mg KOH/g oil. These characteristics were unique to the light crude oil and were essential in understanding its behavior and evaluating the performance of the TEPA Imidazoline inhibitor in effectively controlling corrosion within the pipeline.

The oil pipeline itself had specific specifications and properties that influenced the evaluation. It had a length of 100 km, a diameter of 36 in., and was constructed using steel with a specified grade API 5L X65. The pipeline operated at a designated pressure of 800 psi and transported oil over a distance of 500 km. The oil flow rate in the pipeline was maintained at a specific rate, typically around 1000–2000 bpd, corresponding to a flow velocity of approximately 2–4 m/s. These values were chosen to represent the typical operating conditions of the pipeline. Furthermore, the flow regime within the pipeline was predominantly turbulent. This was attributed to the pipeline's diameter, which ranged from 10 to 20 in., and the properties of the transported oil. The turbulent flow regime ensured effective mixing and distribution of the oil, enhancing heat transfer, and minimizing the potential for localized corrosion. The turbulence was promoted by the high flow velocities and the viscosity of the oil.

The selected steel for the pipeline construction had a chemical composition with approximately 0.2–0.3% carbon (C), 1.0–1.5% manganese (Mn), 0.1–0.3% silicon (Si), and trace amounts of other alloying elements. In terms of mechanical properties, the steel exhibited a tensile strength of around 550–650 megapascals (MPa) and a yield strength of approximately 450–550 MPa. These values were carefully chosen to ensure the steel's ability to withstand the stresses and pressures experienced during pipeline operation. Additionally, the steel possessed excellent corrosion resistance characteristics suitable for the pipeline's operating conditions. It was designed to have a low corrosion rate, typically less than 0.1 millimeters per year (mm/year), when exposed to the corrosive environment of the transported oil and the surrounding atmosphere.

The TEPA Imidazoline inhibitor, which was the focus of the study, had its own distinct properties. The molecular structure of TEPA Imidazoline consisted of four ethylenediamine branches attached to a central pentamine core, forming a complex molecular framework. The chemical composition of TEPA Imidazoline involved elements such as carbon (C), hydrogen (H), nitrogen (N), and potentially other trace elements depending on the manufacturing process. The inhibitor exhibited a solubility of approximately 10% (w/w) in water, allowing for effective dispersion and distribution within the test solutions. In terms of density, TEPA Imidazoline had a value of around 0.9 g/cm^3, indicating a relatively low density compared to water. It demonstrated a high reactivity with the steel surface, forming a protective film or layer when in contact with the steel. This reactivity contributed to its corrosion control capabilities and enhanced its ability to inhibit corrosion processes. These properties played a crucial role in determining its effectiveness as a corrosion inhibitor. To accurately assess the performance of the inhibitor, different concentrations of the inhibitor were used in both the electrochemical and spectroscopic experiments. Concentration ranges from 0.1% to 1% (v/v) were investigated to determine the optimal dosage for effective corrosion control.

Based on the findings of the study, recommendations were made to ensure long-term reliability and corrosion protection within the oil pipeline. It was advised to apply the TEPA Imidazoline inhibitor at an optimal dosage range of 50–100 ppm using injection systems strategically placed along the pipeline, spaced at regular intervals of approximately every 50–100 km. Additionally, implementation of regular monitoring strategies, such as corrosion monitoring probes, visual inspections, and coupon analysis, was recommended at intervals of three to six months. By following these recommendations, the oil pipeline company can optimize inhibitor dosage, ensure uniform application, and establish a proactive monitoring approach to maintain effective corrosion protection and enhance the pipeline's long-term reliability.

Questions

1 What were the specific electrochemical methods employed in the evaluation of the TEPA Imidazoline inhibitor's corrosion control effectiveness, and how were they utilized?
2 Can you elaborate on the realistic parameters of the spectroscopic methods used to analyze the TEPA Imidazoline inhibitor's chemical bonds and its interaction with the steel surface?
3 What were the concentrations of the TEPA Imidazoline inhibitor used in the electrochemical methods, and how were these concentrations determined?
4 Describe the realistic laboratory environmental conditions that were established to simulate the actual environmental conditions in the oil pipeline.
5 What were the detailed specifications of the oil pipeline, including its diameter, length, and material composition?
6 Provide an overview of the properties and specifications of the steel used in the pipeline construction, including its chemical composition and mechanical properties.
7 How were the operation parameters and conditions of the oil pipeline defined and controlled during the experimental evaluation of the TEPA Imidazoline inhibitor?
8 Explain the physical and chemical properties of the water phase present in the transported oil and its significance in the corrosion control evaluation of the inhibitor.

6.2 Comprehensive Evaluation of an Imidazoline Inhibitor for Corrosion Control in an Oil Pipeline Using Advanced Laboratory Techniques

In this case study, the effectiveness of a candidate imidazoline inhibitor is evaluated for controlling corrosion in an oil pipeline. Corrosion in pipelines can lead to significant economic losses, environmental damage, and safety risks. Therefore, finding effective corrosion control strategies is crucial for maintaining the integrity and longevity of the pipeline infrastructure, of the following specifications:

1) **Pipeline material**: The oil pipeline is constructed using API 5L Grade X52 carbon steel, which offers a balance of strength, toughness, and weldability suitable for pipeline applications.

2) **Pipeline diameter**: The pipeline has a diameter of 24 in. (approximately 610 mm). Pipeline diameter may vary depending on the specific project and transportation requirements.
3) **Pipeline wall thickness**: The wall thickness of the pipeline is 0.375 in. (approximately 9.5 mm). The wall thickness is selected based on factors such as operating pressure, pipeline diameter, and regulatory requirements.
4) **Operating pressure**: The pipeline operates at a pressure of 10 MPa or (1450 psi). The operating pressure may vary depending on specific project, terrain, and transportation needs.
5) **Flow rate**: The oil pipeline is designed to transport crude oil at a flow rate of 100,000 bpd. The flow rate can vary depending on factors such as the oil production rate, market demand, and pipeline capacity.
6) **Operating temperature**: The oil pipeline operates at a temperature of 40 °C (104 °F). The operating temperature may vary depending on the ambient conditions and the characteristics of the transported oil.
7) **Pipeline coating**: The internal surface of the oil pipeline is coated with a corrosion-resistant material, such as an epoxy-based or polyurethane-based coating, to provide a protective barrier against corrosion and ensure the smooth flow of oil. The coating thickness typically ranges from 200 to 500 microns (0.008–0.02 in.), with the specific thickness determined by factors such as operating conditions, expected service life, and regulatory requirements.
8) **Cathodic protection**: The pipeline employs a cathodic protection system to mitigate corrosion. Sacrificial anodes or impressed current systems are utilized to provide a protective electrical potential to the pipeline, preventing corrosion from occurring.
9) **Pigging**: Regular pigging operations are conducted to clean and inspect the pipeline. Pigs, devices inserted into the pipeline, are propelled by the flow of the product to remove debris, scale, and other contaminants, ensuring the efficient and safe operation of the pipeline.
10) **Inspection and maintenance**: The oil pipeline undergoes regular inspection and maintenance activities to identify any potential integrity issues. Techniques such as inline inspection (ILI) using smart pigs and visual inspection are employed to detect corrosion, cracks, or other defects that may compromise the pipeline's structural integrity.

To assess the inhibitor's performance, a comprehensive analysis is conducted using various laboratory equipment and techniques. The study involves replicating the pipeline environment in corrosion testing cells, which provide controlled conditions. Metal samples, made of API 5L Grade X52 carbon steel, are selected as a representative material. The composition of API 5L Grade X52 carbon steel typically includes approximately

0.16–0.22% carbon, 1.40% manganese, 0.45% silicon, 0.025% phosphorus, 0.015% sulfur, and trace amounts of other alloying elements. The composition of API 5L Grade X52 carbon steel provides the steel with desirable mechanical strength, toughness, and weldability. The tensile strength of API 5L Grade X52 carbon steel typically ranges from 455 to 600 MPa. The yield strength of this steel is typically around 360–510 MPa. API 5L Grade X52 carbon steel demonstrates good weldability, allowing for efficient joining of sections or repair work in the pipeline.

In the laboratory experiments, the corrosive solutions used to simulate the conditions found in the oil pipeline contain various chemical species and contaminants commonly encountered in these environments, listed as follows:

1) **Chloride concentration**: The corrosive solution typically contains chloride ions in the range of 100–500 ppm. This concentration represents the chloride levels commonly observed in oil pipelines.
2) **Sulfide concentration**: The solution may include sulfide ions, often in the range of 10–50 ppm. Sulfides can originate from the presence of H_2S gas or sulfur-containing compounds in the transported oil.
3) **pH value**: The pH of the corrosive solution is typically adjusted to represent the acidic or mildly acidic conditions found in oil pipelines. A pH range of 4.5–6.5 is commonly employed to simulate the corrosive environment.
4) **Organic acids**: The solution may contain organic acids, such as acetic acid or formic acid, which can result from the decomposition of organic compounds present in crude oil. Concentrations can range from 10 to 50 ppm, reflecting typical levels found in pipelines.
5) **Dissolved gases**: Corrosive solutions may also incorporate dissolved gases like CO_2 and O_2. CO_2 levels can vary from 10 to 100 ppm, while oxygen may be present in trace amounts due to dissolved oxygen in the water phase.
6) **Solid particles**: In some cases, the corrosive solutions may contain solid particles to mimic the presence of suspended solids encountered in oil pipelines. These particles can have sizes ranging from micrometers to millimeters.

To evaluate the corrosion behavior of the metal samples, an electrochemical workstation is employed, utilizing several techniques. The polarization resistance (R_p) technique involves measuring the resistance of the metal sample to corrosion using a small potential perturbation (e.g., ±10 mV), with the resulting current response used to calculate the R_p. EIS is utilized, applying a small AC signal (e.g., 10 mV) across a range of frequencies (typically 0.01 Hz–100 kHz) to the metal-electrolyte interface. Impedance spectra analysis provides information about corrosion processes, including charge transfer resistance (R_{ct}) and double-layer capacitance (C_{dl}). Potentiodynamic polarization measurements involve sweeping the potential of the metal sample between anodic and cathodic regions at a controlled scan rate (e.g., 1 mV/s). This technique yields polarization curves that provide insights into corrosion potential (E_{corr}), corrosion current density (i_{corr}), and the breakdown potential of the passive film. The corrosion rate is calculated using the Tafel extrapolation method, employing the R_p or potentiodynamic polarization data. Corrosion rates are expressed in millimeters per year (mm/year) or micrometers per year (μm/year), representing metal loss over time.

The effectiveness of the imidazoline inhibitor in reducing corrosion is assessed by performing measurements both with and without its presence in the corrosive solution. By comparing corrosion parameters (corrosion rate, R_p, and impedance spectra) under these two conditions, the inhibitor's efficacy can be determined. A lower corrosion rate, increased R_p, and significant changes in impedance spectra indicate improved corrosion protection provided by the inhibitor.

Additionally, autoclaves or high-pressure vessels are employed to evaluate the inhibitor's performance under elevated pressure conditions. Metal samples, along with the corrosive solution containing the inhibitor, are subjected to high pressure, typically 15 MPa, at the designated temperature. These experiments simulate the actual operating conditions of the pipeline, providing insights into the inhibitor's efficacy in real-world scenarios.

Spectroscopic techniques, including Fourier-transform infrared spectroscopy (FTIR) are used to analyze the molecular structure of the imidazoline inhibitor and its interaction with the metal surfaces. FTIR spectra are recorded before and after exposure to determine any changes in functional groups, indicating the formation of protective films. Gas chromatography-mass spectrometry (GC-MS) analysis is performed to identify and quantify degradation products or by-products formed during the interaction between the inhibitor and corrosive agents. This analysis helps evaluate the inhibitor's stability and compatibility in the pipeline environment, ensuring its long-term effectiveness.

Scanning electron microscope (SEM) coupled with energy dispersive X-ray spectroscopy (EDS) is employed to examine the metal surface morphology before and after exposure to the inhibitor. SEM provides high-resolution images that reveal the presence of corrosion products and the formation of protective films. EDS analysis offers elemental composition information, aiding in understanding the inhibitor's effect on the metal surface and the presence of any protective elements. For instance, SEM-EDS analysis may show a smoother surface and a lower presence of iron oxide (Fe_2O_3) deposits on the metal surface treated with the inhibitor.

Mass loss measurements are conducted by immersing metal samples in the corrosive solution, both with and without the inhibitor, for a specific period, such as 30 days. The weight loss of the samples is measured, and corrosion rates are calculated to determine the inhibitor's effectiveness in reducing corrosion. The mass loss calculation methods follow standardized procedures, ensuring reliable and comparable results. For example, the corrosion rate may decrease from 0.5 mm/year without the inhibitor to 0.2 mm/year with the inhibitor, indicating a significant reduction in corrosion.

pH and conductivity measurements are carried out at regular intervals during the corrosion experiments. This helps monitor changes in solution acidity/basicity and conductivity due to the presence of the inhibitor. Standard pH and conductivity meters are used to ensure accurate measurements. For instance, the pH may increase from 4.5 without the inhibitor to 6.0 with the inhibitor, indicating a shift toward a more alkaline environment. Gas or liquid chromatography techniques are employed to analyze the composition of corrosion products formed on the metal surface. This analysis provides insights into the corrosion mechanisms and the inhibitory effect of the imidazoline inhibitor. By identifying the corrosion products, the effectiveness of the inhibitor in preventing their formation can be assessed. For example, chromatographic analysis may reveal a lower concentration of Fe_2O_3 and other corrosion products on the metal surface treated with the inhibitor. Temperature-controlled baths and ovens are used to maintain specific temperatures throughout the experiments. The metal samples, corrosive solutions, and inhibitor formulations are exposed to different temperatures representative of pipeline conditions, such as ambient temperature (25 °C) and elevated temperatures (e.g., 60 °C). These temperature variations allow for an evaluation of the inhibitor's performance under different thermal conditions.

This case study demonstrates a comprehensive evaluation of a candidate imidazoline inhibitor for corrosion control in an oil pipeline. By employing a range of laboratory equipment and techniques, and various analytical instruments, a thorough assessment of the inhibitor's performance under realistic conditions can be conducted. Adherence to relevant standards and codes, such as ASTM G1 and NACE TM0177, ensures reliable and comparable results. The findings of this study, including values such as corrosion rates, impedance spectra, pH levels, and chromatographic analysis results, will contribute to the development of effective corrosion control strategies for oil pipelines, promoting the safety and longevity of the infrastructure.

Questions

1. What is the specific type of carbon steel used for the metal samples in the corrosion evaluation of the oil pipeline? Provide its composition and mechanical properties.
2. Explain how the electrochemical impedance spectroscopy (EIS) technique can provide information about the corrosion processes of the metal samples in the oil pipeline. What are the key parameters extracted from EIS data, and how are they related to corrosion behavior?
3. Describe the process of calculating the corrosion rate using the Tafel extrapolation method based on the polarization resistance or potentiodynamic polarization data. How is the corrosion rate expressed, and what factors can affect its accuracy?
4. What are the limitations and challenges associated with evaluating the effectiveness of the imidazoline inhibitor in reducing corrosion in oil pipelines using electrochemical techniques? How can these challenges be addressed in experimental studies?
5. Discuss the role of chloride ions in the corrosive solutions used to simulate the conditions in oil pipelines. How do chloride ions contribute to corrosion, and what concentration range is typically encountered in pipeline environments?
6. Explain the significance of the polarization resistance (R_p) measurement in assessing the corrosion behavior of metal samples. How does R_p relate to the resistance of the metal to corrosion, and how can it be determined from the obtained current response?
7. Discuss the potential risks and implications of corrosion in oil pipelines. What are the potential consequences of corrosion-related failures, and how can effective corrosion control strategies mitigate these risks?
8. Compare and contrast the advantages and limitations of potentiodynamic polarization and electrochemical impedance spectroscopy (EIS) techniques in evaluating the corrosion behavior of metal samples. When is each technique preferred, and what specific information can they provide?
9. In addition to the imidazoline inhibitor, what other corrosion control methods or technologies can be employed to mitigate corrosion in oil pipelines? Discuss their principles of operation and their potential synergistic effects with the inhibitor.
10. Elaborate on the importance of standardization and compliance with industry codes and regulations in corrosion evaluation and control for oil pipelines. How do international standards and guidelines ensure reliable and consistent assessment of corrosion behavior and the effectiveness of corrosion inhibitors?

11 Describe the mechanism of action of the imidazoline inhibitor in reducing corrosion in oil pipelines. How does it interact with the metal surface and the corrosive solution to provide corrosion protection?

12 Explain the concept of polarization resistance and its significance in corrosion monitoring. How does polarization resistance relate to the corrosion rate, and what information can be derived from its measurement?

13 Discuss the factors that can influence the accuracy and reliability of corrosion rate calculations using the Tafel extrapolation method. How can potential pitfalls or uncertainties in the extrapolation process be addressed to obtain more accurate corrosion rate values?

14 In the context of electrochemical impedance spectroscopy (EIS), explain the significance of the Nyquist plot and the information it provides about the corrosion processes. How can different features of the Nyquist plot be related to specific corrosion mechanisms?

15 Explore the challenges and considerations involved in conducting corrosion evaluations in real-world oil pipeline environments. How can the laboratory-based experimental techniques and results be effectively translated and applied to actual pipeline conditions, taking into account variables such as flow rates, temperature fluctuations, and varying levels of exposure to corrosive agents?

6.3 Comprehensive Experimental Evaluation of CorroShield™: Laboratory, Flow Loop, and Field Assessments for Corrosion Prevention in an Oil Pipeline

Corrosion poses a significant challenge in the oil and gas industry, particularly in the context of maintaining the integrity and reliability of oil pipelines. The damaging effects of corrosion can lead to safety hazards, environmental risks, and substantial economic losses. To combat this issue, the development and evaluation of effective corrosion inhibitors are paramount. This case study presents the evaluation and implementation of a new imidazoline inhibitor, CorroShield™, as a corrosion prevention solution for an oil pipeline. The focus lies on assessing the inhibitor's efficacy in inhibiting corrosion and prolonging the lifespan of the pipeline. The study incorporates multiple stages of evaluation, including laboratory tests, flow loop simulations, and field trials. CorroShield™, the innovative imidazoline-based corrosion inhibitor, is carefully analyzed in terms of its chemical properties and specifications. The inhibitor's ability to form a protective film on metal surfaces, impeding the corrosion process, is a key aspect of interest. Understanding the inhibitor's performance characteristics is crucial to determining its effectiveness in preventing corrosion in an oil pipeline environment. The evaluation process encompasses rigorous experimentation in controlled laboratory conditions, simulating corrosive environments commonly encountered in oil pipelines. A variety of corrosion testing methods, including electrochemical techniques, are employed to quantify the inhibitor's corrosion inhibition capabilities accurately.

A rotating cylinder electrode (RCE) and a flow loop system were utilized to simulate the operational conditions within an oil pipeline. This provides a more comprehensive evaluation of CorroShield™, as it takes into account factors such as flow rate, temperature, and the presence of various substances within the pipeline. The corrosion rates and performance of the inhibitor are monitored and assessed using designated corrosion coupons placed strategically along the flow loop. To validate the inhibitor's performance under real-world conditions, field trials are conducted within an operational oil pipeline. The pipeline's dimensions, flow characteristics, and the composition of transported substances are taken into consideration. The field evaluation involves employing corrosion monitoring techniques and inspections to assess the effectiveness of CorroShield™ in mitigating corrosion and maintaining the pipeline's integrity.

1) **Experimental evaluation**: Three types of experimental evaluation were conducted to assess the effectiveness of the new imidazoline inhibitor, for corrosion prevention in an oil pipeline. The parameters for each evaluation were as follows:
   a) **Laboratory evaluation**: In the laboratory setting, corrosion tests were performed using a rotating cylinder electrode system. The test conditions included a temperature of 60 °C, a pH of 6.5, and a solution containing 3% sodium chloride (NaCl) to simulate the corrosive environment. The concentration of CorroShield™ used was 500 ppm. The corrosion rate was measured using electrochemical techniques, such as linear polarization resistance (LPR) and EIS. The experiments were conducted for a duration of 30 days.
   b) **Flow loop evaluation**: A flow loop system was constructed to simulate the flow conditions in an oil pipeline. The components of the flow loop included a stainless steel pipe with an inner diameter of 10 cm, a pump capable of achieving a flow rate of 10,000 barrels per day (bpd), and a temperature-controlled environment set to 40 °C. The concentration of CorroShield™ in the pipeline was maintained at 100 ppm. Corrosion coupons made of the new-generation steel were installed at various locations along the pipeline to assess the

corrosion rate. The flow loop experiments were conducted continuously for a period of 90 days.

c) **Field evaluation**: Field evaluation was conducted in an operational oil pipeline with similar specifications as the flow loop system. The pipeline had a diameter of 30 cm and carried an oil/water/gas flow. The flow rate was maintained at 50,000 bpd. CorroShield™ was injected into the pipeline at a concentration of 200 ppm using an injection system. Corrosion monitoring techniques, such as ultrasonic thickness measurements and visual inspections, were employed to assess the effectiveness of the inhibitor. The field evaluation was conducted over a period of one year.

2) **Imidazoline inhibitor**: CorroShield™ is a new-generation imidazoline-based corrosion inhibitor specifically developed for the oil and gas industry. It is a liquid compound with a chemical formula $C_{12}H_{23}N_3O_2$. The inhibitor has a molecular weight of 249.33 g/mol and a density of 0.88 g/cm^3. CorroShield™ is soluble in both water and hydrocarbons, making it suitable for application in oil pipelines. It exhibits excellent film-forming properties and provides a protective barrier on metal surfaces, inhibiting the corrosion process.

3) **Flow loop specifications**:
   a) The flow loop system used in the evaluation of CorroShield™ consisted of the following components and specifications:
   b) **Stainless steel pipe**: Inner diameter of 10 cm, length of 100 m.
   c) **Pump**: Capable of achieving a flow rate of 10,000 bpd.
   d) **Temperature control**: Maintained at a constant 40 °C.
   e) **Corrosion coupons**: Made of the new generation steel, installed at various locations along the pipeline for monitoring purposes.

4) **New generation steel**: The new generation steel used in the oil pipeline possesses excellent corrosion resistance properties. A typical HSLA (High-Strength, Low-Alloy) steel used in oil pipelines would have the following mechanical properties and chemical composition:
   Mechanical properties:
   Yield strength: Typically ranging from 450 to 550 MPa
   Tensile strength: Typically ranging from 550 to 650 MPa
   Elongation: Generally greater than 20%
   Charpy impact toughness: Typically greater than 60 J at −20 °C
   Chemical composition:
   Carbon (C): Typically less than 0.20%
   Manganese (Mn): Generally ranging from 1.00% to 1.50%
   Phosphorus (P): Typically less than 0.03%
   Sulfur (S): Typically less than 0.03%

Silicon (Si): Generally ranging from 0.20% to 0.50%
Nickel (Ni): Generally less than 0.50%
Chromium (Cr): Generally less than 0.50%
Copper (Cu): Generally less than 0.40%
Vanadium (V): Generally ranging from 0.05% to 0.15%

5) **Oil pipeline specifications**: The oil pipeline used in the study had the following specifications:
   Diameter: 30 cm
   Length: Varies depending on the operational requirements
   Medium: Carried an oil/water/gas flow
   Flow rate: Maintained at 50,000 bpd
6) **Oil/water/gas flow**: The oil/water/gas flow in the pipeline had the following physical and chemical properties:
   Oil: Crude oil with a density of 0.85 g/cm^3 and a viscosity of 20 cP
   Water: Stratum water with a density of 1.02 g/cm^3, comprising a mixture of chemical constituents including 2% NaCl, 0.5% calcium carbonate ($CaCO_3$), 0.3% magnesium sulfate ($MgSO_4$), and 0.1% potassium chloride (KCl).Gas: Natural gas with a pressure of 5 bar and a composition of 90% methane (CH_4) and 10% ethane (C_2H_6)
   The flow rate of the oil/water/gas mixture was maintained at a constant 50,000 bpd.
7) **Standards and codes**: The evaluation and implementation of CorroShield™ for corrosion prevention in the oil pipeline adhered to the following relevant standards and codes:
   NACE International Standard SP0176-2021: "Corrosion Control of Steel Fixed Offshore Platforms Associated with Petroleum Production."
   API Recommended Practice 571: "Damage Mechanisms Affecting Fixed Equipment in the Refining Industry."
   ASTM International Standard G1: ""Standard Practice for Preparing, Cleaning, and Evaluating Corrosion Test Specimens."
8) **Recommendations for pipeline operators, and corrosion prevention strategy**: Based on the evaluation results, the following recommendations are suggested for pipeline operators:
   a) A regular corrosion monitoring program shall be implemented using techniques such as ultrasonic thickness measurements and visual inspections.
   b) The inhibitor concentration shall be continuously monitored, and the injection rate shall be adjusted accordingly to ensure optimum corrosion protection.
   c) Periodic maintenance activities, such as inhibitor replenishment and cleaning of the pipeline, shall be conducted.

d) Personnel shall be trained and educated on corrosion prevention strategies and the proper handling and application of CorroShield™.

By implementing these recommendations, pipeline operators can effectively mitigate corrosion issues and ensure the long-term integrity of the oil pipeline, thereby enhancing its operational reliability and reducing maintenance costs.

9) **Corrosion prevention strategy**: The corrosion prevention strategy mentioned above involves the application of a corrosion inhibitor called CorroShield™. This inhibitor is used to protect the metal surface of the steel used in oil pipelines from corrosion. It is applied in two different concentrations: 100 ppm in the flow loop and 200 ppm in the field evaluation. CorroShield™ works by forming a protective film on the surface of the steel. This film acts as a barrier between the metal and the corrosive environment, preventing or slowing down the corrosion process. By forming this protective layer, CorroShield™ helps to extend the service life of the steel by reducing the rate of corrosion and minimizing the damage caused by corrosive elements such as moisture, chemicals, and other corrosive agents present in the pipeline environment. The concentration of 100 ppm in the flow loop and 200 ppm in the field evaluation refers to the amount of CorroShield™ added to the pipeline system. Parts per million (ppm) is a unit of concentration that indicates the ratio of the amount of CorroShield™ to the total volume of the solution or medium. In this case, the inhibitor is added in relatively low concentrations to ensure its effectiveness in forming the protective film while also maintaining cost-effectiveness. By applying CorroShield™ and forming a protective film on the metal surface, the corrosion prevention strategy aims to reduce the occurrence of corrosion and its detrimental effects, ultimately enhancing the longevity and performance of the steel used in the oil pipeline.

Questions

1 How does the corrosion rate of the new generation steel in the laboratory evaluation compare to the corrosion rate in the flow loop and field evaluations? What factors might contribute to any variations observed?

2 Considering the chemical composition of the saline water in the flow loop, how does the presence of multiple chemical constituents affect the effectiveness of CorroShield™ in preventing corrosion? Are there any specific interactions or reactions that could influence its performance?

3 In the field evaluation, were there any observed differences in the inhibitor's performance in different sections of the oil pipeline? What factors,

such as varying flow rates or composition of the oil/water/gas mixture, might contribute to these differences?

4 How does the concentration of CorroShield™ used in the laboratory evaluation compare to the concentrations employed in the flow loop and field evaluations? Were there any significant differences in corrosion inhibition observed at different inhibitor concentrations?
5 What are the key limitations or challenges associated with implementing CorroShield™ as a corrosion inhibitor in an operational oil pipeline? Are there any specific operational conditions or factors that may impact the inhibitor's long-term effectiveness?
6 In the laboratory evaluation, what other electrochemical techniques, besides linear polarization resistance (LPR) and electrochemical impedance spectroscopy (EIS), were employed to evaluate the corrosion rate of the new generation steel? How did these additional techniques contribute to the overall assessment?
7 Can the performance of CorroShield™ in preventing corrosion be influenced by external factors such as temperature fluctuations, pressure variations, or the presence of other additives in the oil/water/gas flow? How were these factors addressed or controlled in the experimental evaluations?
8 How do the findings and recommendations of this case study align with existing industry standards and codes, such as NACE International Standard SP0176-2021 and API Recommended Practice 571? Are there any additional considerations or modifications to existing standards that should be considered based on the evaluation results?

6.4 Comparative Evaluation of Four Steel Types and Imidazoline-based Corrosion Inhibitor for Oil Pipeline Protection

The oil pipeline under consideration is designed to transport crude oil from an offshore production facility to an onshore refinery. The pipeline has a length of 300 km and a diameter of 24 in. It is constructed using carbon steel due to its cost-effectiveness and adequate mechanical properties for oil transportation. The pipeline is buried underground to minimize external factors that could lead to corrosion. It is equipped with cathodic protection systems to provide additional corrosion prevention. The multiphase flow in the pipeline consists of crude oil, natural gas, and water. The flow rate varies depending on the production levels, but on average, the oil flow rate is 10,000 bpd. The oil transported through the pipeline is a complex mixture primarily composed of hydrocarbons and sulfur compounds. The hydrocarbons, including alkanes, aromatics, cycloalkanes, and olefins, constitute approximately 70% of the crude oil by weight. Alkanes make up the largest fraction, accounting for 40–60% of the total crude oil. Aromatics contribute to about 10–30%, while cycloalkanes and olefins are present in concentrations ranging from 5% to 15% and around 5% or less, respectively. Sulfur compounds, notably H_2S and various organic sulfur compounds, are also prevalent in the crude oil. H_2S, a highly corrosive and toxic gas, can be present in concentrations ranging from trace amounts up to several thousand ppm. Organic sulfur compounds, such as mercaptans and sulfides, are typically found in the range of a few ppm to a few hundred ppm.

1) Four different steel types were to be examined with the following details:
   a) **ASTM A106 Grade B carbon steel**: ASTM A106 Grade B carbon steel, a commonly utilized type in pipeline applications, exhibits a specific chemical composition to meet the required mechanical properties and corrosion resistance. It typically contains a maximum carbon content of 0.30%, ensuring a balance between strength and weldability. Manganese, ranging from 0.29% to 1.06%, enhances the steel's tensile strength and overall toughness. The maximum allowable phosphorus and sulfur contents are 0.035%, minimizing their detrimental effects on weldability and corrosion resistance. The minimum silicon content of 0.10% improves the steel's deoxidation properties, while copper, nickel, chromium, molybdenum, and vanadium, present within specified limits (0.40% maximum for each), contribute to the steel's strength, corrosion resistance, and other desirable properties. ASTM A106 Grade B carbon steel exhibits favorable mechanical properties and a distinct microstructure suitable for pipeline applications. The steel demonstrates a minimum tensile strength of 415 MPa (60,000 psi) and a minimum yield strength of 240 MPa (35,000 psi), providing excellent structural integrity and load-bearing capabilities. Its elongation upon fracture is 30%, indicating good ductility and the ability to withstand deformation without brittle failure. The microstructure of ASTM A106 Grade B carbon steel typically consists of ferrite and pearlite phases. Ferrite, a relatively soft and ductile phase, contributes to the steel's toughness and impact resistance. Pearlite, a mixture of ferrite and cementite, provides strength and hardness to the steel. This microstructure, along with the specified mechanical properties, ensures the ASTM A106 Grade B carbon steel is well-suited for the demanding conditions of pipeline environments.

b) **AISI 316L stainless steel**: AISI 316L stainless steel, a common type employed in pipeline applications, possesses a specific chemical composition that imparts its corrosion-resistant properties. It typically consists of approximately 16–18% chromium, which forms a protective chromium oxide layer on the surface, guarding against corrosion and oxidation. The steel also contains 10–14% nickel, which enhances its resistance to both general and localized corrosion. Additionally, AISI 316L stainless steel includes 2–3% molybdenum, which further enhances its corrosion resistance, especially in chloride-rich environments. The composition is typically low in carbon, with a maximum of 0.03%, to prevent sensitization and maintain the steel's corrosion resistance. These chemical constituents, along with other minor elements such as manganese, silicon, phosphorus, sulfur, and nitrogen within specified limits, contribute to the superior corrosion-resistant properties of AISI 316L stainless steel, making it well-suited for pipeline applications in challenging and corrosive environments. AISI 316L stainless steel exhibits excellent mechanical properties and a distinct microstructure that make it suitable for pipeline applications requiring corrosion resistance. It has a high tensile strength of approximately 515 MPa (75,000 psi) and a yield strength of around 205 MPa (30,000 psi), cnsuring structural integrity and durability under applied loads. With an elongation of 40% or more, AISI 316L stainless steel demonstrates good ductility and the ability to undergo deformation without fracturing. The microstructure of AISI 316L stainless steel primarily consists of austenite, a face-centered cubic (FCC) crystal structure that provides high toughness and excellent corrosion resistance. This microstructure is achieved through the addition of nickel and molybdenum, which stabilize the austenitic structure.

c) **UNS S32205, also known as Duplex 2205 stainless steel**: UNS S32205, commonly referred to as Duplex 2205, is a duplex stainless steel that exhibits a specific chemical composition, enabling it to provide exceptional corrosion resistance and mechanical properties. It typically contains approximately 22% chromium, which contributes to its excellent resistance against various corrosive environments, including acids, chloride ions, and seawater. The steel also consists of around 5–6% nickel, enhancing its toughness and resistance to stress corrosion cracking. Additionally, Duplex 2205 contains approximately 3% molybdenum, which further enhances its resistance to pitting and crevice corrosion. The composition includes approximately 0.03% carbon, ensuring a balanced combination of corrosion resistance and weldability. With the addition of small amounts of other elements such as manganese, silicon, phosphorus, sulfur, nitrogen, and a trace of copper, Duplex 2205 achieves a chemical composition that allows it to excel in demanding pipeline applications where resistance to corrosion and mechanical strength are paramount. Duplex 2205 stainless steel exhibits excellent mechanical properties and a unique microstructure that makes it highly suitable for pipeline applications. It offers a high tensile strength of approximately 620 MPa (90,000 psi) and a yield strength of around 450 MPa (65,000 psi), providing superior structural integrity and load-bearing capabilities. With a minimum elongation of 25%, Duplex 2205 stainless steel demonstrates good ductility and the ability to withstand deformation without failure. The microstructure of Duplex 2205 stainless steel is a balanced combination of ferrite and austenite phases, commonly referred to as a duplex structure. This microstructure is achieved through its chemical composition, including a higher content of chromium, nickel, and nitrogen. The duplex microstructure provides exceptional corrosion resistance, high strength, and improved resistance to stress corrosion cracking compared to conventional stainless steels. It also exhibits good toughness, making it resistant to brittle fracture in harsh operating conditions.

d) **AISI 4140 alloy steel**: AISI 4140 alloy steel is a widely utilized type of alloy steel in pipeline applications, renowned for its excellent strength and toughness properties. Its chemical composition typically includes approximately 0.38–0.43% carbon, providing a good balance of strength and machinability. The steel contains approximately 0.75–1% chromium, which contributes to its hardenability and corrosion resistance. Manganese content ranges from 0.75 to 1.00%, enhancing the steel's hardenability and strength. Additionally, AISI 4140 alloy steel contains approximately 0.15–0.25% molybdenum, improving its high-temperature strength and creep resistance. Silicon, phosphorus, and sulfur are present in small amounts to enhance specific properties, while traces of other elements, such as nickel and copper, may also be present. This chemical composition, combined with appropriate heat treatment, imparts AISI 4140 alloy steel with superior mechanical properties, including high tensile strength, excellent toughness, and good wear resistance, making it well-suited for demanding pipeline applications. AISI 4140 alloy steel possesses impressive mechanical properties and a distinct microstructure that make it highly suitable for pipeline applications requiring strength and durability. It exhibits a high tensile strength of approximately 850 MPa (120,000 psi) and a yield strength of around 680 MPa (100,000 psi), providing excellent structural integrity and load-bearing capabilities. AISI 4140 alloy steel also demonstrates good toughness, allowing it to absorb energy and withstand impact without fracturing. The microstructure of AISI 4140 alloy steel is predominantly martensitic, resulting from the heat treatment process. This microstructure provides the steel with its high strength and hardness. In certain conditions, tempered martensite or bainite microstructures can also be achieved to balance strength and ductility.

2) **The test inhibitor**: Fatty acid imidazoline (FAI) inhibitor. It is typically available as a liquid formulation with a moderate viscosity, facilitating convenient handling, mixing, and application. The inhibitor's chemical composition includes a balance of fatty acid chain lengths, usually derived from C12 to C18 fatty acids, optimizing its hydrophobicity and solubility. It forms a protective film on the metal surface, preventing the access of corrosive agents and mitigating corrosion. Its density, fall within a typical range of 0.9–1.1 g/cm^3. It is commonly available as a liquid with a moderate viscosity, falling within a range of 100–500 cP. The solubility in hydrocarbon-based fluids is excellent, allowing for efficient dispersion throughout the pipeline system. The flash point of FAI is typically above 100 °C (212 °F), ensuring safe handling and storage.
3) The relationship between the four different steels and the inhibition performance with the use of electrochemical methods. The electrochemical methods were conducted using a simulated crude oil electrolyte with a 3% water content, creating an environment that closely represents the corrosive conditions typically encountered in oil pipelines. To simulate real-world conditions, the pH of the electrolyte was adjusted accordingly. The experiments were carried out at a controlled temperature of 25 °C to ensure consistency and allow for accurate comparison of the results. During the electrochemical measurements, a R_p scan rate of 0.1 mV/s was employed to analyze the corrosion behavior of the different steel types. Additionally, EIS was performed within a frequency range of 10 mHz–100 kHz to evaluate the electrochemical response and provide insights into the inhibition performance of the steel samples. The selected parameters and conditions allowed for a comprehensive assessment of the relationship between the four different steels and the inhibition performance, shedding light on their corrosion resistance and suitability for oil pipeline applications. Also, the surface was examined with scanning electron microscopy

4) **Results**:
   a) **ASTM A106 Grade B carbon steel**: Without the inhibitor, the corrosion rate was measured at 2.5 mpy, and extensive corrosion products and surface pitting were observed. With the addition of the inhibitor, the corrosion rate decreased to 0.2 mpy, indicating significant inhibition, and the steel surface showed a significantly reduced amount of corrosion products and pitting, indicating effective protection.
   b) **AISI 316L stainless steel**: The corrosion rate without the inhibitor was 1.2 mpy, localized corrosion and surface etching were visible. With the application of the inhibitor, the corrosion rate reduced to 0.1 mpy, and the surface showed fewer signs of localized corrosion exhibiting effective inhibition.
   c) **UNS S32205 (or Duplex 2205) stainless steel**: In the absence of the inhibitor, the corrosion rate measured 0.8 mpy, with slight signs of localized corrosion observed. With the addition of the inhibitor, the corrosion rate dropped to 0.05 mpy, and the steel surface exhibited no significant signs of corrosion, indicating excellent inhibition performance.
   d) **AISI 4140 alloy steel**: Without the inhibitor, the corrosion rate was 3.0 mpy, and visible corrosion attacks and surface roughening were observed. After the addition of the inhibitor, the corrosion rate decreased to 0.3 mpy, and the steel surface showed reduced corrosion attacks and improved surface condition, indicating good inhibition.

   The nature of the corrosion products was analyzed using techniques such as XRD and EDS. The analysis revealed that the corrosion products primarily consisted of Fe_2O_3 and iron sulfide (FeS), indicating the presence of both oxygen and sulfur corrosion mechanisms.
5) **Recommendation of the best steel type**: Based on the comprehensive evaluation, including mechanical properties, corrosion rates, and inhibition performance, the duplex stainless steel demonstrated the best overall performance. It exhibited high strength, good corrosion resistance, and excellent inhibition when treated with the inhibitor. Therefore, the recommendation is to utilize duplex stainless steel for the oil pipeline construction.
6) **Recommendation of the inhibitor use in the field**: Considering the positive results obtained from the electrochemical and microscopic studies, it is recommended to use fatty acid imidazoline inhibitor as a corrosion inhibitor in the field for the oil pipeline. Its inhibitive properties, stability, and compatibility with crude oil make it suitable for long-term corrosion protection. Proper dosage and application methods should be

determined based on pipeline specifications and operational conditions to ensure optimal inhibition performance. Regular monitoring and maintenance should also be implemented to ensure continued inhibitor effectiveness.

Questions

1 How do the corrosion rates of the four different steel types compare when subjected to the same simulated crude oil electrolyte conditions, and how does this information influence the selection of the most corrosion-resistant steel for oil pipeline applications?
2 What are the specific impedance values obtained from the electrochemical impedance spectroscopy (EIS) measurements for each steel type at different frequencies, and how do these values correlate with their inhibition performance?
3 What is the influence of the imidazoline inhibitor on the electrochemical properties of the steel samples? Does the inhibitor demonstrate a significant decrease in corrosion rates or an increase in polarization resistance for all steel types?
4 How does the chemical composition and microstructure of each steel type contribute to its corrosion resistance and performance in the presence of the imidazoline inhibitor?
5 Can the results obtained from the electrochemical methods be validated and complemented by other corrosion evaluation techniques, such as weight loss measurements or corrosion product analysis, to provide a more comprehensive understanding of the corrosion behavior and inhibition performance of the steel types?

6.5 Laboratory Evaluation of Octadecylamine-based Imidazoline Inhibitor for Corrosion Rate Reduction in Alloy Steel Used for an Oil Pipeline Using XPS, AFM, STM, PM-IRRAS, EDX, XRD, and FTIR

In this case study, a comprehensive laboratory evaluation is conducted on an octadecylamine-based imidazoline inhibitor to examine its efficiency in reducing the corrosion rate of an alloy steel used for an oil pipeline. The evaluation focused on utilizing various advanced laboratory techniques to analyze the inhibitor's impact on the alloy steel surface at different concentrations. These techniques included X-ray photoelectron spectroscopy (XPS), atomic force microscopy characterization (AFM), scanning tunneling microscope (STM), polarization modulation infrared reflection absorption spectroscopy (PM-IRRAS), energy-dispersive X-ray spectroscopy (EDX), X-Ray diffraction (XRD), and Fourier-transform infrared spectroscopy (FTIR). By employing these methods, important insights were gained into the inhibitor's ability to modify the alloy steel surface, form a protective film, inhibit corrosion reactions, and alter the chemical composition of the surface. The results of this laboratory evaluation provide valuable data on the effectiveness of the octadecylamine-based imidazoline inhibitor, enabling informed decisions for its potential application in the oil industry to enhance the corrosion resistance and longevity of oil pipelines.

1) **X-ray photoelectron spectroscopy (XPS)**: XPS analysis was conducted to evaluate the efficiency of an octadecylamine-based imidazoline inhibitor in reducing the corrosion rate of an alloy steel used for an oil pipeline. Five different concentrations of the inhibitor were tested, ranging from 0.1% to 1% by weight. The XPS results revealed valuable insights into the surface composition and chemical bonding states of the alloy steel samples. The XPS spectra exhibited characteristic peaks corresponding to the alloy steel surface elements, such as iron (Fe) and chromium (Cr), as well as the inhibitor compound. The relative intensities of these peaks varied with the concentration of the inhibitor. At higher inhibitor concentrations, a noticeable decrease in the intensity of Fe and Cr peaks was observed, indicating a reduction in the surface content of these elements. This reduction can be attributed to the formation of a protective film by the inhibitor on the alloy steel surface. Furthermore, the binding energies of Fe and Cr species were analyzed, providing information on their chemical states. In the absence of the inhibitor, Fe and Cr were predominantly present in the form of oxides, suggesting ongoing corrosion processes. However, as the inhibitor concentration increased, the binding energy of Fe and Cr shifted toward lower values, indicating the formation of passivating species. This shift confirmed the effectiveness of the octadecylamine-based imidazoline inhibitor in inhibiting the corrosion of the alloy steel.
2) **Atomic force microscopy (AFM) characterization**: The AFM technique was employed to investigate the surface morphology and topography of the alloy steel samples treated with various concentrations of the octadecylamine-based imidazoline inhibitor. The AFM images provided valuable insights into the surface roughness and the formation of a protective film on the alloy steel surface. The AFM analysis revealed that the untreated alloy steel surface exhibited a relatively high roughness, indicating the presence of corrosion products and surface irregularities.

However, as the concentration of the inhibitor increased, a noticeable decrease in surface roughness was observed. This reduction in roughness indicated the formation of a smoother surface, suggesting the effective inhibition of corrosion processes. Furthermore, the AFM images displayed the formation of a uniform and compact film on the alloy steel surface as the inhibitor concentration increased. The thickness of the protective film increased proportionally with the inhibitor concentration, which confirmed the film-forming properties of the octadecylamine-based imidazoline inhibitor. These findings demonstrated the ability of the inhibitor to provide a barrier against corrosion and protect the alloy steel surface.

3) **Scanning tunneling microscope (STM)**: STM analysis was employed to examine the surface topography and atomic-scale characteristics of the alloy steel samples treated with varying concentrations of the octadecylamine-based imidazoline inhibitor. The STM images provided valuable information regarding the formation of a protective film and the morphology of the alloy steel surface. The STM observations revealed that the untreated alloy steel surface exhibited irregularities, such as pits and surface defects, which are indicative of corrosion damage. However, with the addition of the inhibitor, a remarkable improvement in surface morphology was observed. The STM images displayed a smoother surface with a reduced density of surface defects, indicating the inhibitory effect of the inhibitor on corrosion. Additionally, the STM images confirmed the formation of a self-assembled monolayer (SAM) on the alloy steel surface as the inhibitor concentration increased. The SAM appeared as a well-organized layer of inhibitor molecules, tightly bound to the metal surface. This molecular arrangement contributed to the inhibition of corrosive species' access to the alloy steel surface, effectively reducing the corrosion rate. The STM analysis provided direct evidence of the inhibitor's ability to modify the alloy steel surface and mitigate corrosion processes.
4) **Polarization modulation infrared reflection-absorption spectroscopy (PM-IRRAS)**: The PM-IRRAS technique was employed to analyze the chemical composition and molecular structure of the protective film formed on the alloy steel surface treated with varying concentrations of the octadecylamine-based imidazoline inhibitor. The PM-IRRAS measurements provided valuable information regarding the chemical interactions between the inhibitor molecules and the alloy steel surface. The PM-IRRAS spectra displayed characteristic absorption bands corresponding to the inhibitor compound, indicating its presence on the alloy steel surface. The intensity of these absorption bands increased with the

inhibitor concentration, suggesting a higher coverage of inhibitor molecules on the surface. The presence of absorption peaks corresponding to functional groups of the inhibitor confirmed its adsorption and chemisorption onto the alloy steel surface. Moreover, the PM-IRRAS analysis provided information on the structural changes occurring within the inhibitor molecules upon adsorption. The shift in peak positions and changes in peak intensities indicated the formation of coordination bonds and the rearrangement of molecular configurations. These structural changes were responsible for the inhibitor's ability to form a protective film and hinder the corrosion process.

5) **Energy-dispersive X-ray spectroscopy (EDX)**: The EDX technique was employed to investigate the elemental composition and distribution on the surface of the alloy steel samples treated with various concentrations of the octadecylamine-based imidazoline inhibitor. The EDX analysis provided valuable information regarding the formation of the protective film and its elemental constituents. The EDX spectra revealed the presence of iron (Fe), chromium (Cr), and other alloying elements on the untreated alloy steel surface. However, as the inhibitor concentration increased, a reduction in the relative intensity of Fe and Cr peaks was observed, indicating the formation of a protective film enriched with the inhibitor compound. This reduction in elemental content confirmed the inhibitor's ability to inhibit the corrosion of the alloy steel. Furthermore, the EDX mapping analysis displayed a uniform distribution of the inhibitor elements across the alloy steel surface treated with higher inhibitor concentrations. This homogeneous distribution indicated the formation of a continuous and well-adhered protective film, ensuring effective corrosion inhibition throughout the exposed surface area. The EDX results provided strong evidence supporting the protective nature of the octadecylamine-based imidazoline inhibitor.
6) **X-ray diffraction (XRD)**: XRD analysis was conducted to examine the crystallographic structure and phase composition of the alloy steel samples treated with different concentrations of the octadecylamine-based imidazoline inhibitor. The XRD measurements provided valuable information regarding the formation of corrosion products and the structural changes induced by the inhibitor. The XRD patterns obtained from the untreated alloy steel samples exhibited characteristic peaks corresponding to various iron oxide phases, indicating ongoing corrosion processes. However, as the inhibitor concentration increased, a significant decrease in the intensity of these peaks was observed, suggesting a suppression of corrosion and reduced formation of oxide phases. Moreover, the XRD analysis revealed the presence of additional peaks corresponding to the

inhibitor compound in the treated alloy steel samples. These peaks confirmed the deposition of the inhibitor onto the alloy steel surface and the formation of a protective film. The absence of new peaks corresponding to other corrosion products further supported the inhibitor's effectiveness in reducing the corrosion rate of the alloy steel.

7) **Fourier-transform infrared spectroscopy (FTIR)**: The FTIR analysis was employed to investigate the molecular interactions and chemical bonding between the octadecylamine-based imidazoline inhibitor and the alloy steel surface. The FTIR spectra provided valuable information on the inhibitor's adsorption behavior and the formation of a protective film. The FTIR spectra exhibited characteristic absorption bands corresponding to the functional groups present in the inhibitor compound. With increasing inhibitor concentration, the intensity of these absorption bands increased, indicating a higher coverage of inhibitor molecules on the alloy steel surface. Shifts in peak positions and changes in peak shapes suggested the formation of chemical bonds between the inhibitor and the alloy steel surface, promoting the formation of a stable and protective film. Furthermore, the FTIR analysis provided insights into the interaction between the inhibitor and the alloy steel surface functional groups. Changes in the intensity and position of absorption bands corresponding to surface functional groups indicated the formation of inhibitor-surface complexes. These complexes played a crucial role in inhibiting the corrosion process by forming a barrier and preventing corrosive species' access to the alloy steel surface.

Questions

1 How does the X-ray photoelectron spectroscopy (XPS) analysis provide insights into the efficiency of the octadecylamine-based imidazoline inhibitor in reducing the corrosion rate of the alloy steel?

2 What information can be obtained from the wire beam electrode (WBE) technique regarding the electrochemical behavior and corrosion resistance of the alloy steel in the presence of the inhibitor?

3 How does atomic force microscopy (AFM) characterization contribute to the understanding of the surface morphology and topography of the alloy steel samples treated with different concentrations of the inhibitor?

4 Explain how the scanning tunneling microscope (STM) analysis provides valuable insights into the surface topography and atomic-scale characteristics of the alloy steel samples treated with the octadecylamine-based imidazoline inhibitor.

5 How does polarization modulation infrared reflection-absorption spectroscopy (PM-IRRAS) help in understanding the chemical interactions between the inhibitor and the alloy steel surface, as well as the formation of a protective film?
6 In what way does energy-dispersive X-ray spectroscopy (EDX) contribute to the evaluation of the octadecylamine-based imidazoline inhibitor's effectiveness in reducing the corrosion rate of the alloy steel?
7 How does X-ray diffraction (XRD) analysis provide information about the crystallographic structure and phase composition of the alloy steel samples treated with different concentrations of the inhibitor?
8 Explain the role of Fourier-transform infrared spectroscopy (FTIR) in investigating the molecular interactions and chemical bonding between the inhibitor and the alloy steel surface.
9 How does the wire beam electrode (WBE) technique provide information on the corrosion current density and passivation behavior of the alloy steel samples treated with the octadecylamine-based imidazoline inhibitor?
10 Describe the insights gained from the atomic force microscopy (AFM) analysis regarding the formation of a protective film and the reduction in surface roughness of the alloy steel samples treated with the inhibitor.

7

Case Studies: Field Operations

7.1 Comprehensive Protection and Performance Evaluation of an Oil Pipeline Using DBSI Corrosion Inhibitor

XYZ Company operates an extensive oil pipeline network for transporting crude oil from oilfields to refineries. To ensure the pipeline's integrity and mitigate corrosion, XYZ Company has implemented a comprehensive protection and monitoring program using dodecylbenzene sulfonic acid imidazoline (DBSI) as the corrosion inhibitor. This case study outlines the pipeline's key parameters, operation conditions, relevant codes, and the procedures for protecting, monitoring, and evaluating the performance of the corrosion inhibitor.

Pipeline description: The oil pipeline is a 24-in. diameter carbon steel pipeline with a length of 500 miles. The pipeline operates at a pressure of 900 psi and transports crude oil containing small amounts of dissolved oxygen and hydrogen sulfide (H_2S). The steel used in the pipeline construction conforms to API 5L Grade X70.

Corrosion inhibitor: DBSI, a well-established corrosion inhibitor, has been selected for this pipeline. It is known for its excellent film-forming properties and effective corrosion inhibition against both oxygen and H_2S. DBSI is compatible with carbon steel and has a proven track record in protecting oil pipelines from corrosion.

Imidazoline Inhibitors for Corrosion Protection of Oil Pipeline Steels: Experimental Laboratory Evaluation and Case Studies, First Edition. Faysal Fayez Eliyan.
© 2024 John Wiley & Sons, Inc. Published 2024 by John Wiley & Sons, Inc.

Relevant Codes:

- **API 570**: Piping Inspection Code
- **NACE SP0169-2013**: Control of External Corrosion on Underground or Submerged Metallic Piping Systems
- **NACE MR0175/ISO 15156**: Petroleum and Natural Gas Industries – Materials for Use in H_2S-containing Environments in Oil and Gas Production

Procedure for Protecting and Monitoring the Oil Pipeline:

1) **Injection system installation**: The DBSI injection equipment, including pumps, flow meters, and control valves, is installed at strategic locations along the pipeline. Compatibility with DBSI and capability of providing accurate dosage control are ensured for the equipment. This installation process ensures that the injection system is in place and ready to deliver the corrosion inhibitor efficiently.
2) **Injection rate calculation**: To determine the required DBSI injection rate, various pipeline parameters such as diameter, flow rate, inhibitor concentration, and desired dosage are taken into account. The appropriate injection rate is calculated by following the recommendations and guidelines provided by the DBSI manufacturer. This calculation ensures that the inhibitor is injected at the correct rate to effectively protect against corrosion.
3) **DBSI solution preparation**: The DBSI corrosion inhibitor solution is prepared by diluting the DBSI concentrate with a suitable diluent according to the manufacturer's guidelines. Thorough mixing is ensured to achieve a homogeneous solution. This preparation process guarantees that the corrosion inhibitor is properly mixed and ready for injection.
4) **Injection monitoring and control**: The injection system parameters, including injection rate, inhibitor concentration, and pipeline pressure, are continuously monitored. Regular calibration of the injection equipment is performed to maintain accurate dosage control. Monitoring and control measures are implemented to ensure that the injection process is functioning correctly and delivering the corrosion inhibitor as intended.
5) **Routine inspections and maintenance**: Periodic inspections of the pipeline are conducted to identify any signs of corrosion or coating degradation. Regular maintenance of the injection equipment is performed to ensure its proper functioning. These inspections and maintenance activities help to detect and address any potential issues promptly, ensuring the ongoing effectiveness of the corrosion protection system.

Procedure for Evaluating the Performance of the Inhibitor:

1) **Corrosion rate monitoring**: Regular corrosion rate measurements are conducted at specific locations along the pipeline using corrosion probes or other suitable techniques. The measured corrosion rates are compared with pre-established baseline values to evaluate the inhibitor's performance. This monitoring process allows for the assessment of the corrosion protection provided by the inhibitor.
2) **Metal loss analysis**: Pipeline sections are periodically inspected for metal loss using non-destructive testing methods such as ultrasonic thickness measurements. The results of these inspections are analyzed to assess the level of corrosion protection provided by the inhibitor. Metal loss analysis helps to identify any areas where the inhibitor may not be adequately protecting the pipeline.
3) **Coating integrity assessment**: The pipeline's external coating is inspected for any signs of degradation, such as cracking or blistering. The coating's integrity is assessed to ensure it provides an additional layer of protection against corrosion. Coating integrity assessment helps to identify any vulnerabilities in the protective coating that may compromise corrosion protection.
4) **Environmental monitoring**: The pipeline's environment, including temperature, humidity, and soil conditions, is monitored to identify any factors that may affect corrosion rates or inhibitor performance. Environmental monitoring allows for the identification of external factors that may impact the corrosion protection system's effectiveness.
5) **Performance evaluation and adjustment**: The collected data from corrosion rate monitoring, metal loss analysis, coating integrity assessment, and environmental monitoring is evaluated and compared against industry standards and internal performance targets. Based on this evaluation, adjustments may be made to the DBSI injection rate, inhibitor concentration, or injection locations as necessary to optimize corrosion protection. Performance evaluation and adjustment ensure that the corrosion protection system remains effective and meets the required standards.

Questions

1 How does the DBSI corrosion inhibitor effectively mitigate both oxygen and hydrogen sulfide (H_2S) corrosion in the presence of dissolved oxygen and trace amounts of H_2S in the crude oil?

2 What are the specific industry standards or codes mentioned in the case study that guide the protection and evaluation of the oil pipeline? How do these standards ensure compliance and optimal corrosion control?

3 Describe the methodology and techniques used to measure the corrosion rate along the pipeline. How do these measurements assist in evaluating the performance of the DBSI corrosion inhibitor?
4 How is the injection process monitored and controlled to ensure accurate dosage and proper delivery of the corrosion inhibitor?
5 What measures are taken to evaluate the performance of the DBSI corrosion inhibitor, and how are adjustments made to optimize corrosion protection based on the evaluation results?
6 Considering the 500-mile length of the pipeline, what challenges or considerations should XYZ Company address when implementing routine inspections and maintenance activities? How can these challenges be effectively managed to ensure the long-term integrity of the pipeline and the continued effectiveness of the DBSI corrosion inhibitor?

7.2 Effective Protection and Performance Evaluation of an Oil Pipeline Using Octadecylamine Imidazoline (ODA) Corrosion Inhibitor

XYZ Company operates a critical oil pipeline network, which is susceptible to corrosion caused by the presence of highly corrosive water along with oil. The pipeline's integrity is safeguarded through a comprehensive protection and monitoring program that utilizes octadecylamine imidazoline (ODA) as the corrosion inhibitor. This case study highlights the pipeline's key parameters, significance of flow regime, operation conditions, relevant codes, procedures for protecting and monitoring the pipeline, and the evaluation of ODA inhibitor performance.

Pipeline description: The oil pipeline is a 16-in. diameter carbon steel pipeline, spanning 200 miles. The pipeline carries a mixture of oil and highly corrosive water. The steel used in the pipeline is API 5L Grade X65. The pipeline operates at an average pressure of 600 psi, with varying flow rates depending on demand and seasonal factors.

Corrosion inhibitor: ODA, a widely used corrosion inhibitor, has been selected to protect the pipeline from corrosion caused by the corrosive water-oil mixture. ODA is known for its excellent film-forming properties and effectiveness in mitigating both general and localized corrosion. It is compatible with carbon steel and has demonstrated successful results in similar oil pipeline applications.

Significance of flow regime: The flow regime within the pipeline plays a crucial role in the corrosion process. Transitions between laminar and turbulent flow can impact the corrosion rate. Additionally, the presence of stagnant zones or regions of low flow can promote localized corrosion. Understanding and managing the flow regime is essential for effective corrosion control.

Relevant Codes:

- **API 570**: Piping Inspection Code
- **NACE SP0169-2013**: Control of External Corrosion on Underground or Submerged Metallic Piping Systems
- **NACE MR0175/ISO 15156**: Petroleum and natural gas industries – Materials for use in H_2S-containing environments in oil and gas production

The procedure for protecting and monitoring the oil pipeline includes several steps, which are outlined below.

1) **Injection system installation**: The ODA injection equipment, including pumps, flow meters, and control valves, is installed at strategic locations along the pipeline. The compatibility of the equipment with ODA and its capability to provide accurate dosage control are ensured.
2) **Injection rate calculation**: The required ODA injection rate is calculated based on pipeline parameters such as diameter, flow rate, inhibitor concentration, and desired dosage. The appropriate injection rate is determined following the recommendations and guidelines provided by the ODA manufacturer.
3) **ODA solution preparation**: The ODA corrosion inhibitor solution is prepared by diluting the ODA concentrate with a suitable diluent, following the manufacturer's guidelines. Thorough mixing is done to achieve a homogeneous solution.
4) **Injection monitoring and control**: The injection system parameters, including injection rate, inhibitor concentration, pipeline pressure, and flow regime, are continuously monitored. The injection equipment is regularly calibrated to maintain accurate dosage control.
5) **Routine inspections and maintenance**: Periodic inspections of the pipeline are conducted to identify any signs of corrosion or coating degradation. Regular maintenance of the injection equipment is performed to ensure its proper functioning.

Procedure for Evaluating the Performance of the Inhibitor:

The procedure for evaluating the performance of the inhibitor involves the following steps.

1) **Corrosion rate monitoring**: The corrosion rate is regularly measured at selected locations along the pipeline using techniques such as corrosion probes or electrochemical methods. The measured corrosion rates are compared with pre-established baseline values to evaluate the inhibitor's performance.
2) **Coating integrity assessment**: The external coating is inspected for any signs of degradation, including coating thickness, adhesion, and coverage. The integrity of the coating is assessed to ensure it provides an additional layer of protection against corrosion.
3) **Material analysis**: Samples are collected from the pipeline's internal surface and analyzed using techniques like scanning electron microscopy (SEM) and energy-dispersive X-ray spectroscopy (EDS). The samples are evaluated for signs of corrosion, pitting, or inhibitor deposition to assess the inhibitor's performance and identify potential areas of concern.
4) **Environmental monitoring**: The pipeline's environment, including water chemistry, pH, temperature, and flow characteristics, is monitored. Variations or anomalies that may impact corrosion rates or inhibitor effectiveness are identified.
5) **Performance evaluation and adjustments**: The collected data is analyzed and compared against industry standards and internal performance targets. Areas for improvement are identified, and adjustments to the ODA injection rate, inhibitor concentration, or injection locations are made as needed to optimize corrosion protection.

Questions

1 How does the presence of highly corrosive water in the oil pipeline impact the probability of failure? Discuss the specific corrosion mechanisms involved and their potential consequences for the pipeline's integrity.

2 Explain the significance of flow regime in relation to corrosion control in the oil pipeline.

3 Compare the performance and effectiveness of octadecylamine imidazoline (ODA) corrosion inhibitor in mitigating general corrosion and localized corrosion in the presence of highly corrosive water. Discuss

the underlying mechanisms that make ODA suitable for these corrosion types.

4 Describe the challenges involved in maintaining accurate dosage control of the ODA corrosion inhibitor, considering the varying flow rates and pressures in the oil pipeline. How can XYZ Company ensure precise and consistent injection of the inhibitor under these conditions?

5 Analyze the potential interactions between the ODA corrosion inhibitor and the highly corrosive water–oil mixture. What factors should XYZ Company consider to ensure the compatibility and stability of the inhibitor in the presence of such complex fluid compositions?

7.3 Mitigation of Stress Corrosion Cracking in an Oil Pipeline using Stearoylamine Imidazoline (SAI) Corrosion Inhibitor

XYZ Company operates a critical oil pipeline network and aims to prevent stress corrosion cracking (SCC) through an effective protection and monitoring program. This case study focuses on the pipeline's key parameters, the significance of coating, the utilization of stearoylamine imidazoline (SAI) as the corrosion inhibitor, operation conditions, relevant codes, and the procedures for protecting, monitoring, and evaluating the performance of the inhibitor.

Pipeline description: The oil pipeline is constructed with 20-in. diameter carbon steel, spanning a distance of 300 miles. The pipeline operates at an average pressure of 800 psi and carries crude oil. The steel used in the pipeline construction adheres to API 5L Grade X60.

Corrosion inhibitor: SAI has been chosen as the corrosion inhibitor for mitigating SCC in the pipeline. SAI is known for its ability to form a protective film on the metal surface, reducing the possibility of SCC initiation and propagation. It demonstrates compatibility with carbon steel and has a proven track record in similar oil pipeline applications.

Significance of coating: The pipeline's external coating plays a critical role in preventing corrosion by acting as a physical barrier between the corrosive environment and the steel substrate. A well-maintained coating provides additional protection against environmental factors and enhances the performance of the corrosion inhibitor.

Relevant Codes:

- **API 570**: Piping Inspection Code
- **NACE SP0169-2013**: Control of External Corrosion on Underground or Submerged Metallic Piping Systems
- **NACE TM0185**: Evaluation of Pipeline and Pressure Vessel Steels for Resistance to Hydrogen-Induced Cracking

Procedure for Protecting and Monitoring the Oil Pipeline:

1) **Coating application and inspection**: Apply a suitable external coating to the pipeline surface to provide a protective barrier against corrosion. Regularly inspect the coating for any signs of damage, such as cracks, peeling, or blistering, and repair any coating defects promptly.
2) **SAI injection system installation**: Install SAI injection equipment, including pumps, flow meters, and control valves, at strategic locations along the pipeline. Ensure the equipment is compatible with SAI and capable of providing accurate dosage control.
3) **Injection rate calculation**: Calculate the required SAI injection rate based on pipeline parameters such as diameter, flow rate, inhibitor concentration, and desired dosage. Follow SAI manufacturer's guidelines and industry best practices to determine the appropriate injection rate.
4) **SAI solution preparation**: Prepare the SAI corrosion inhibitor solution by diluting the SAI concentrate with a suitable diluent as per the manufacturer's recommendations. Ensure proper mixing to achieve a homogenous solution.
5) **Injection monitoring and control**: Continuously monitor the injection system parameters, including injection rate, inhibitor concentration, pipeline pressure, and temperature. Regularly calibrate the injection equipment to maintain accurate dosage control.

Procedure for Evaluating the Performance of the Inhibitor:

1) **SCC monitoring**: Implement regular inspections using non-destructive testing techniques such as ultrasonic testing (UT) or magnetic particle inspection (MPI) to detect any signs of SCC. Assess the severity and extent of SCC and compare with baseline data to evaluate the effectiveness of the SAI corrosion inhibitor.
2) **Coating integrity assessment**: Conduct periodic inspections to evaluate the integrity of the pipeline's external coating. Inspect for coating defects, such as disbondment or holidays, which may compromise the protective barrier and increase the risk of SCC.

3) **Material analysis**: Collect representative samples from the pipeline's internal surface to analyze the microstructure and assess the presence of SCC or other corrosion-related phenomena. Utilize techniques like metallography, scanning electron microscopy (SEM), and energy-dispersive X-ray spectroscopy (EDS) to investigate the material properties and corrosion mechanisms.
4) **Environmental monitoring**: Continuously monitor the pipeline's environment, including water chemistry, pH, temperature, and the presence of corrosive contaminants. Evaluate any changes in the environmental conditions that may impact the performance of the SAI inhibitor and increase the risk of SCC.
5) **Performance evaluation and adjustments**: Analyze the collected data, comparing it against industry standards, performance targets, and established baseline values. Adjust the SAI injection rate, inhibitor concentration, or injection locations as necessary to optimize SCC protection and minimize the risk of failure.

Questions

1. Describe the underlying mechanisms by which stearoylamine imidazoline (SAI) corrosion inhibitor effectively reduces the possibility of stress corrosion cracking (SCC) in the oil pipeline. How does SAI mitigate the interactions between the corrosive environment and the steel substrate?
2. Discuss the challenges associated with maintaining the integrity of the pipeline's external coating and its significance in preventing stress corrosion cracking. How does the coating interact with the SAI inhibitor to provide enhanced protection against SCC?
3. Explain the specific parameters used in calculating the required SAI injection rate. How do variations in flow rate, inhibitor concentration, and pipeline diameter affect the dosage control and overall effectiveness of the corrosion inhibitor in preventing SCC?
4. In the evaluation of the SAI inhibitor's performance, describe the techniques and tools used to detect and assess stress corrosion cracking. How can non-destructive testing methods such as UT or MPI provide insights into the presence and severity of SCC?
5. Considering the operation conditions of the oil pipeline, discuss the potential interactions between the corrosive environment, the SAI inhibitor, and the steel substrate. How can XYZ Company effectively monitor and adapt the inhibitor's performance to mitigate the risks associated with SCC in dynamic operating conditions?

7.4 Suppression of CO_2/H_2S Corrosion in an Oil Pipeline using Laurylamine Imidazoline (LAI) Inhibitor

XYZ Company operates a critical oil pipeline network that faces the challenge of CO_2/H_2S corrosion due to the presence of carbon dioxide (CO_2) and H_2S in the transported oil. This case study focuses on the effectiveness of laurylamine imidazoline (LAI) inhibitor in conjunction with other inhibitors to suppress CO_2/H_2S corrosion. It also considers the significance of steel type, operation parameters, operation conditions, relevant codes, and procedures for protecting, monitoring, and evaluating the performance of the inhibitor.

Pipeline description: The oil pipeline is constructed using various steel types, including API 5L Grade X52 and X70, to withstand the demanding operating conditions. The pipeline has a diameter of 24 in. and extends over a distance of 400 miles. It operates at an average pressure of 900 psi, transporting oil containing significant amounts of CO_2 and H_2S.

Effectiveness of LAI inhibitor in suppressing CO_2/H_2S corrosion: LAI inhibitor is used in combination with other inhibitors to provide effective corrosion control in the presence of CO_2 and H_2S. The synergistic effect of these inhibitors helps suppress the corrosive action of CO_2 and H_2S gases and reduce the risk of corrosion-related failures.

Significance of steel type: The choice of steel type is crucial in combating CO_2/H_2S corrosion. Steel types with higher alloy content, such as API 5L Grade X70, offer improved resistance to corrosion and stress cracking in the presence of corrosive gases. Understanding the steel's properties and its compatibility with the inhibitors is essential for optimizing corrosion control strategies.

Parameters: Parameters such as pipeline diameter (24 in.), operating pressure (900 psi), flow rate (measured in barrels per day or cubic meters per hour), inhibitor concentration (measured in parts per million), and temperature (measured in degrees Celsius or Fahrenheit) are considered utilized throughout the case study.

Operation conditions: The pipeline operates under challenging conditions with varying oil compositions, flow rates, temperatures, and exposure to CO_2 and H_2S gases. These conditions reflect the realistic challenges faced by XYZ Company and require effective corrosion control strategies.

Relevant Codes:

- **NACE MR0175/ISO 15156**: Petroleum and natural gas industries – Materials for use in H_2S-containing environments in oil and gas production

- **API 570**: Piping Inspection Code
- **NACE SP0204-2017**: Stress Corrosion Cracking (SCC) Direct Assessment Methodology

Procedure for Protecting and Monitoring the Oil Pipeline:

1) **Inhibitor injection system design**: An inhibitor injection system is designed that accommodates the specific pipeline parameters, including diameter, flow rate, and inhibitor concentration. Appropriate pumps, flow meters, control valves, and injection points are incorporated to ensure accurate dosage control and distribution.
2) **Injection rate calculation**: The required injection rate of LAI inhibitor and other inhibitors is calculated based on the pipeline's flow rate, inhibitor concentration, and desired dosage. The compatibility of different inhibitors and their synergistic effects in suppressing CO_2/H_2S corrosion is considered.
3) **Inhibitor solution preparation**: The LAI inhibitor solution is prepared, considering the appropriate concentration and compatibility with other inhibitors used in the corrosion control program. Proper mixing and homogeneity of the inhibitor solution are ensured to maintain consistent performance.
4) **Injection monitoring and control**: The injection system parameters, including injection rate, inhibitor concentration, pipeline pressure, and temperature, are continuously monitored. The injection equipment is regularly calibrated, and the injection rate is adjusted as necessary to maintain optimal corrosion control.
5) **Routine inspection and monitoring**: A comprehensive inspection and monitoring program is implemented to assess the pipeline's condition and corrosion performance. Visual inspections, internal and external corrosion monitoring, and thickness measurements using techniques like UT and corrosion coupons are performed.

Procedure for Evaluating the Performance of the Inhibitor:

1) **Corrosion rate monitoring**: Corrosion rate data is collected using techniques such as corrosion coupons, electrical resistance probes, or linear polarization resistance measurements. Corrosion rates are compared with established industry standards and internal performance targets to evaluate the effectiveness of the inhibitor.
2) **Material analysis**: Periodic material analysis is conducted on pipeline samples to assess the presence of corrosion products, pitting, or other forms of localized corrosion. Techniques like metallography, SEM, and

EDS are utilized to gain insights into the inhibitor's ability to mitigate CO_2/H_2S corrosion.

3) **Coating integrity assessment**: The pipeline's external coating is regularly inspected for any signs of damage or degradation. The coating's integrity and effectiveness in providing an additional barrier against CO_2/H_2S corrosion are assessed.
4) **Environmental monitoring**: The pipeline's environment, including CO_2 and H_2S concentrations, temperature, pH, and water chemistry, is continuously monitored. Any variations or anomalies that may impact the inhibitor's performance and the corrosion rate are evaluated.
5) **Performance evaluation and adjustments**: The collected data is analyzed and compared against industry standards, regulatory requirements, and established performance targets. Adjustments to the inhibitor injection rate, inhibitor concentration, or corrosion control strategies are made as needed to optimize corrosion protection and reduce the possibility of CO_2/H_2S corrosion.

Questions

1. How does the combination of laurylamine imidazoline (LAI) inhibitor with other inhibitors effectively suppress CO_2/H_2S corrosion in the oil pipeline? Explain the synergistic effects and mechanisms involved in mitigating the corrosive action of CO_2 and H_2S gases.
2. Discuss the significance of the steel type, specifically API 5L Grade X52 and X70, in combating CO_2/H_2S corrosion. How do the different steel types offer varying levels of resistance to corrosive gases and contribute to the overall effectiveness of the corrosion control strategy?
3. Describe the detailed procedure and parameters used for protecting and monitoring the oil pipeline against CO_2/H_2S corrosion. How does XYZ Company ensure accurate dosage control, proper injection distribution, and comprehensive inspection and monitoring to optimize corrosion protection?
4. Explain the parameters used in evaluating the performance of the LAI inhibitor in suppressing CO_2/H_2S corrosion. How are corrosion rates measured, material analysis conducted, and coating integrity assessed to assess the effectiveness of the inhibitor in mitigating corrosion risks?
5. In the case of CO_2/H_2S corrosion, how does XYZ Company adapt their corrosion control strategy based on the results of performance evaluation? Describe the process of making adjustments to the inhibitor injection rate, concentration, or corrosion control strategies to optimize protection and minimize the risk of CO_2/H_2S corrosion-induced failures.

7.5 Effective Corrosion Control in a High-Temperature, High-Flow, and High-Pressure Oil Pipeline: A Case Study on the Use of Palmitoylamine Imidazoline (PAI) Inhibitor

XYZ Company operates a critical oil pipeline that faces significant challenges due to high temperature, high flow rate, high pressure, and the presence of saline water. This case study focuses on the effectiveness of palmitoylamine imidazoline (PAI) as a corrosion inhibitor, the significance of steel type, operation parameters, operation conditions, relevant codes, and procedures for protecting, monitoring, and evaluating the performance of the inhibitor.

Pipeline description: The oil pipeline has a diameter of 30 in. and spans a distance of 500 miles. It operates at a high temperature of 150 °C (302 °F), a high flow rate of 100,000 bpd, and a high pressure of 1500 psi. The water present in the pipeline is saline, adding to the corrosion challenges.

Effectiveness of PAI inhibitor: PAI is employed as a corrosion inhibitor to mitigate the corrosive effects of high-temperature and high-salinity environments. Its effectiveness lies in its ability to form a protective film on the metal surface, inhibiting the corrosive reactions and preventing the formation of scales and deposits.

Significance of steel type: The choice of steel type is crucial in maintaining the pipeline's integrity under challenging conditions. High-temperature and high-pressure environments require steel with excellent heat resistance and high tensile strength, such as API 5L Grade X80 or X100, to withstand the stresses and prevent deformation or failure.

Parameters:

1) **Pipeline diameter**: 30 in. (76.2 cm)
2) **Operating temperature**: 150 °C (302 °F)
3) **Flow rate**: 100,000 bpd (13.4 m^3/s)
4) **Pressure**: 1500 psi (10.3 MPa)
5) **Saline water concentration**: 3% (by weight)

Operation conditions: The pipeline operates under challenging conditions characterized by high temperatures, high flow rates, high pressures, and the presence of saline water. These conditions simulate the realistic challenges faced by XYZ Company and require effective corrosion control strategies.

Relevant Codes:

- **NACE SP0176**: Corrosion Control of Submerged Areas
- **API 5L [American Petroleum Institute, 2018]**: Specification for Line Pipe
- **NACE MR0175/ISO 15156**: Petroleum and Natural Gas Industries – Materials for Use in H_2S-Containing Environments

Procedure for Protecting and Monitoring the Oil Pipeline:

1) **Inhibitor compatibility assessment**: Compatibility tests are conducted to ensure the effectiveness of PAI in the specific environment, considering factors such as high temperature, high salinity, and high flow rates. The compatibility of PAI with other inhibitors, if applicable, is evaluated to optimize the corrosion control program.
2) **Injection system design**: An inhibitor injection system is designed that takes into account the pipeline's diameter, flow rate, and pressure. Suitable injection pumps, flow meters, control valves, and injection points are incorporated to ensure accurate dosage control and distribution.
3) **Injection rate calculation**: The required injection rate of PAI is calculated based on the flow rate, desired inhibitor concentration, and the corrosivity of the environment. The effects of temperature and salinity on the inhibitor's performance are considered, and the injection rate is adjusted accordingly.
4) **Inhibitor solution preparation**: The PAI inhibitor solution is prepared with the appropriate concentration and compatibility with other inhibitors, if applicable. Thorough mixing and homogeneity of the inhibitor solution are ensured before injection.
5) **Injection monitoring and control**: The injection system parameters, including injection rate, inhibitor concentration, pipeline pressure, and temperature, are continuously monitored. The injection equipment is regularly calibrated, and the injection rate is adjusted to maintain optimal corrosion control.
6) **Routine inspection and monitoring**: A comprehensive inspection and monitoring program is implemented to assess the pipeline's condition and corrosion performance. Visual inspections, internal and external corrosion monitoring, and thickness measurements using techniques such as UT and corrosion coupons are performed.

Procedure for Evaluating the Performance of the Inhibitor:

1) **Corrosion rate monitoring**: Corrosion rate data is collected using techniques such as corrosion coupons, electrical resistance probes, or

linear polarization resistance measurements. The corrosion rates are compared with industry standards and internal performance targets to evaluate the effectiveness of the PAI inhibitor.

2) **Material analysis**: Periodic material analysis is conducted on pipeline samples to assess the presence of corrosion products, pitting, or other localized forms of corrosion. Techniques like metallography, SEM, and EDS are utilized to gain insights into the inhibitor's ability to mitigate corrosion in high-temperature and high-salinity environments.
3) **Coating integrity assessment**: The pipeline's external coating is regularly inspected for any signs of damage or degradation. The integrity and effectiveness of the coating in providing an additional barrier against corrosion under high-temperature, high-flow, and high-pressure conditions are assessed.
4) **Environmental monitoring**: The pipeline's environment, including temperature, pressure, salinity, pH, and water chemistry, is continuously monitored. Any variations or anomalies that may impact the PAI inhibitor's performance and corrosion rate are evaluated.
5) **Performance evaluation and adjustments**: The collected data is analyzed and compared against industry standards, regulatory requirements, and established performance targets. Adjustments are made to the inhibitor injection rate, concentration, or corrosion control strategies as needed to optimize corrosion protection and minimize the risk of failures.

Questions

1 How does the palmitoylamine imidazoline (PAI) inhibitor effectively mitigate corrosion in a high-temperature, high-flow, and high-pressure oil pipeline? Discuss the specific mechanisms and interactions that enable PAI to provide effective protection in such challenging operating conditions.

2 Explain the significance of steel type, specifically API 5L Grade X80 or X100, in ensuring the integrity of the oil pipeline under high-temperature, high-flow, and high-pressure conditions. How does the choice of steel type contribute to corrosion resistance and prevent structural failures?

3 Considering the high temperature of 150 °C (302 °F) in the oil pipeline, discuss the impact of thermal degradation on the performance of the PAI inhibitor. How does the inhibitor maintain its effectiveness and protective properties under prolonged exposure to high temperatures?

4 Discuss the challenges associated with maintaining the desired inhibitor concentration and injection rate in a high-flow pipeline with a flow rate of 100,000 barrels per day (bpd). How does XYZ Company ensure accurate dosage control and effective distribution of the PAI inhibitor throughout the pipeline?

5 Saline water is present in the pipeline, adding to the corrosion challenges. Explain how the PAI inhibitor addresses the corrosive effects of saline water and its ability to protect the pipeline's integrity under these conditions. What specific properties of PAI make it suitable for combating corrosion in a high-salinity environment?

7.6 Enhancing Corrosion Control in a High H_2S Concentration, High Water Content, and Slug Flow Oil Pipeline: A Case Study on the Dual Use of PAI and EOI Inhibitors

Introduction: XYZ Company operates a critical oil pipeline facing unique challenges, including high H_2S concentration, high water content, and predominantly slug flow regime. This case study focuses on the significance of preventive maintenance, the effectiveness of dual inhibitors (palmitoylamine imidazoline – PAI and ethoxylated imidazoline – EOI), operation parameters, operation conditions, relevant codes and standards, and procedures for protecting, monitoring, and evaluating the performance of the inhibitors.

Pipeline description: The oil pipeline spans a distance of 200 miles, with a diameter of 24 in. and operates under high H_2S concentration and high water content. The flow regime is predominantly slug flow, characterized by intermittent high-velocity slug sections separated by liquid slugs.

Significance of preventive maintenance: Implementing a robust preventive maintenance program is crucial to ensure the pipeline's integrity and minimize the risk of corrosion-related failures. Regular inspections, cleaning, and maintenance activities help identify potential issues early on, allowing for timely repairs and corrosion control measures.

Effectiveness of dual use of PAI and EOI inhibitors: The combined use of PAI and EOI inhibitors offers enhanced corrosion protection in the challenging operating conditions of high H_2S concentration, high water content, and slug flow regime. The inhibitors form a synergistic protective film on the pipeline's internal surface, minimizing the corrosive effects of H_2S and inhibiting the formation of scales and deposits.

Parameters:

1) **Pipeline diameter**: 24 in. (61 cm)
2) **H_2S concentration**: 1000 ppm (parts per million)

3) **Water content**: 10% (by volume)
4) **Slug flow regime**: Intermittent high-velocity slug sections separated by liquid slugs

Operation Conditions:

1) **Operating temperature**: 50 °C (122 °F)
2) **Flow rate**: 30,000 barrels per day (bpd) (4 m^3/s)
3) **Operating pressure**: 500 psi (3.45 MPa)
4) **Material of construction**: API 5L Grade X60 Carbon Steel

Relevant Codes and Standards:

1) **NACE MR0175/ISO 15156**: Petroleum and Natural Gas Industries – Materials for Use in H_2S-Containing Environments
2) **API RP 571**: Damage Mechanisms Affecting Fixed Equipment in the Refining Industry
3) **API RP 574**: Inspection Practices for Piping System Components

Procedure for Protecting and Monitoring the Oil Pipeline:

1) **Pre-commissioning activities**: A thorough pipeline cleaning and pre-commissioning flushing will be conducted to remove debris, mill scale, and construction residues. The pipeline will be ensured to be dry and free of contaminants before the inhibitors are introduced.
2) **Inhibitor injection system design**: An inhibitor injection system will be designed, taking into consideration the pipeline's diameter, flow rate, and pressure. Suitable injection pumps, flow meters, control valves, and injection points will be installed to ensure accurate dosage control and distribution.
3) **Inhibitor dosage calculation**: The required dosage of PAI and EOI inhibitors will be calculated based on the pipeline's flow rate, H_2S concentration, and water content. The corrosivity of the environment will be accounted for, and the dosage will be adjusted to achieve optimal corrosion control.
4) **Injection and distribution**: The injection rates of PAI and EOI inhibitors will be monitored to ensure consistent and adequate distribution throughout the pipeline. Injection flow profiling techniques will be implemented to optimize inhibitor distribution in the slug flow regime.
5) **Regular inspection and monitoring**: Routine visual inspections, internal and external corrosion monitoring, and thickness measurements will be conducted using techniques like UT and corrosion coupons. The H_2S concentration, water content, and flow characteristics

will be monitored to assess the inhibitor performance and identify potential corrosion hotspots.

Procedure for Evaluating the Performance of Inhibitor:

1) **Corrosion rate monitoring**: Corrosion rate data will be collected using techniques such as corrosion coupons, electrical resistance probes, or linear polarization resistance measurements. The corrosion rates will be compared with industry standards and internal performance targets to evaluate the effectiveness of the PAI and EOI inhibitors.
2) **Material analysis**: Pipeline samples will be periodically analyzed to assess the presence of corrosion products, pitting, or other localized forms of corrosion. Techniques like metallography, SEM, and EDS will be utilized to gain insights into the inhibitors' ability to mitigate corrosion in high H_2S concentration and high-water content environments.
3) **Coating integrity assessment**: The pipeline's external coating will be regularly inspected for any signs of damage or degradation. The coating's integrity and effectiveness in providing an additional barrier against corrosion in the slug flow regime will be assessed.
4) **Environmental monitoring**: Continuous monitoring of H_2S concentration, water content, flow regime, and other relevant parameters will be conducted to evaluate the inhibitors' performance and identify any deviations or anomalies.
5) **Performance evaluation and adjustments**: The collected data will be analyzed and compared against industry standards, regulatory requirements, and established performance targets. Adjustments to the inhibitor dosages, injection rates, or corrosion control strategies will be made as needed to optimize corrosion protection and prevent failures.

Questions

1 How does the dual use of palmitoylamine imidazoline (PAI) and ethoxylated imidazoline (EOI) inhibitors effectively mitigate corrosion in an oil pipeline operating under high H_2S concentration, high water content, and slug flow regime? Discuss the synergistic effects and mechanisms by which these inhibitors provide enhanced protection in such challenging conditions.

2 What are the specific challenges posed by high H_2S concentration and high water content in the pipeline, and how do the PAI and EOI inhibitors address these challenges? Discuss the inhibitory properties of these inhibitors against H_2S-induced corrosion and their ability to control corrosion in the presence of high water content.

3 Slug flow regime is known to create complex flow patterns and varying velocities within the pipeline. How do the PAI and EOI inhibitors effectively distribute and adhere to the internal surface of the pipeline under slug flow conditions? Explain the mechanisms by which these inhibitors can overcome the challenges posed by slug flow and provide consistent corrosion protection.
4 What are the limitations or considerations to be aware of when using the PAI and EOI inhibitors in a pipeline with high H_2S concentration and high water content? Discuss factors such as inhibitor compatibility, dosage requirements, and potential interactions with other pipeline components or fluids.
5 How does the implementation of preventive maintenance practices contribute to the overall effectiveness of corrosion control in the pipeline? Discuss the importance of regular inspections, cleaning, and maintenance activities in identifying and mitigating corrosion-related issues, particularly in the context of operating conditions with high H_2S concentration, high water content, and slug flow regime.

7.7 Corrosion Control and Cathodic Protection in a Low-Temperature, Stratified Flow Oil Pipeline: A Case Study on the Effectiveness of Laurylamine Imidazoline (LAI) Inhibitor

Introduction: XYZ Company operates an oil pipeline facing unique challenges, including low temperature, stratified flow regime, and the need for effective corrosion control. This case study focuses on the significance of cathodic protection, the effectiveness of laurylamine imidazoline (LAI) inhibitor, parameters, operation conditions, cathodic protection parameters, relevant codes and standards, and procedures for protecting, monitoring, and evaluating the performance of the inhibitor.

Pipeline description: The oil pipeline spans a distance of 200 miles, with a diameter of 16 in., and operates under low temperature conditions with a stratified flow regime. The pipeline transports oil through various sections with different flow patterns, characterized by the separation of oil and gas phases due to density differences.

Significance of cathodic protection: Cathodic protection is crucial to mitigate corrosion in the pipeline by providing a sacrificial or impressed current to counteract the electrochemical reactions responsible for corrosion. It helps prevent the formation of corrosion cells and protects vulnerable areas, such as welds and pipe joints, from accelerated corrosion.

Effectiveness of laurylamine imidazoline (LAI) inhibitor: The utilization of LAI inhibitor enhances corrosion control in the challenging operating conditions of low temperature and stratified flow. LAI forms a protective film on the pipeline's internal surface, inhibiting the corrosion process and reducing the likelihood of corrosion-related failures.

Parameters:

1) **Pipeline diameter**: 16 in. (40.64 cm)
2) **Operating temperature**: −10 °C (14 °F)
3) **Stratified flow regime**: Separation of oil and gas phases due to density differences

Operation Conditions:

1) **Flow rate**: 10,000 barrels per day (bpd) (1.59 m^3/s)
2) **Operating pressure**: 800 psi (5.52 MPa)
3) **Material of construction**: API 5L Grade X52 Carbon Steel

Cathodic Protection Parameters:

1) **Sacrificial anode system**: Aluminum alloy anodes with a specified weight and surface area distribution along the pipeline.
2) **Impressed current system**: Titanium mixed metal oxide (MMO) anodes, rectifiers, and reference electrodes.
3) **Current density**: Design current density based on the specific requirements of the pipeline, coating condition, and soil resistivity.

Relevant Codes and Standards:

1) **NACE SP0169**: Control of External Corrosion on Underground or Submerged Metallic Piping Systems
2) **API RP 651**: Cathodic Protection of Aboveground Petroleum Storage Tanks
3) **ASTM G57**: Standard Test Method for Field Measurement of Soil Resistivity Using the Wenner Four-Electrode Method

Procedure for Protecting and Monitoring the Oil Pipeline:

1) **Cathodic protection design**: A comprehensive corrosion risk assessment will be conducted and a cathodic protection system will be designed based on industry standards, pipeline parameters, and corrosion control objectives. The appropriate type of cathodic protection system (sacrificial or impressed current) will be considered based on the characteristics of the pipeline and expected corrosion threats.

2) **Cathodic protection installation**: Sacrificial anodes or impressed current anodes will be installed along the pipeline at specified intervals. Proper electrical connections, backfill materials, and bonding between pipeline sections will be ensured to optimize the performance of the cathodic protection system.
3) **LAI inhibitor application**: The required dosage of LAI inhibitor will be calculated based on the pipeline's flow rate, temperature, and expected corrosion rate. The inhibitor will be injected into the pipeline at designated injection points to ensure uniform distribution and adherence to the internal surface.
4) **Cathodic protection monitoring**: The pipeline's cathodic protection system will be regularly monitored using techniques such as potential measurements, current mapping, and close-interval potential surveys (CIPS). The performance of the anodes, current density, and potential readings will be measured and recorded to evaluate the effectiveness of the cathodic protection system.
5) **Corrosion monitoring**: Routine visual inspections, internal and external corrosion monitoring, and thickness measurements will be conducted using techniques like UT and corrosion coupons. Corrosion data and trends will be analyzed to identify potential corrosion hotspots and take necessary corrective actions.

Procedure for Evaluating the Performance of LAI Inhibitor:

1) **Corrosion rate monitoring**: Corrosion rate data will be collected using techniques such as corrosion coupons, electrical resistance probes, or linear polarization resistance measurements. The corrosion rates will be compared with industry standards and internal performance targets to evaluate the effectiveness of the LAI inhibitor in mitigating corrosion.
2) **Material analysis**: Pipeline samples will be periodically analyzed to assess the presence of corrosion products, pitting, or other localized forms of corrosion. Techniques such as metallography, SEM, and EDS will be utilized to gain insights into the inhibitor's ability to mitigate corrosion in low-temperature and stratified flow conditions.
3) **Coating integrity assessment**: The pipeline's external coating will be regularly inspected for any signs of damage or degradation. The integrity and effectiveness of the coating in providing an additional barrier against corrosion in the specific operating conditions will be assessed.
4) **Environmental monitoring**: Continuous monitoring of H_2S concentration, water content, flow regime, and other relevant parameters will

be carried out to evaluate the inhibitor's performance and identify any deviations or anomalies.

5) **Performance evaluation and adjustments**: The collected data will be analyzed and compared against industry standards, regulatory requirements, and established performance targets. Adjustments to the LAI inhibitor dosage, injection rates, or corrosion control strategies will be made as needed to optimize corrosion protection and prevent failures.

Questions

1 How does the utilization of cathodic protection and the effectiveness of laurylamine imidazoline (LAI) inhibitor contribute to mitigating corrosion in a low-temperature environment with a stratified flow regime? Discuss the synergistic effects and mechanisms by which these corrosion control measures provide enhanced protection in such challenging conditions.

2 What are the specific challenges posed by low temperature and stratified flow regime in the pipeline, and how do the cathodic protection system and LAI inhibitor address these challenges? Discuss the inhibitory properties of the LAI inhibitor against corrosion and its ability to provide effective protection in the presence of stratified flow.

3 How does the cathodic protection system effectively distribute protective currents along the pipeline's internal surface under the conditions of low temperature and stratified flow? Discuss the design considerations and parameters involved in optimizing the performance of the cathodic protection system for corrosion control in this unique operating environment.

4 What are the limitations or considerations to be aware of when using LAI inhibitor in a pipeline with low-temperature and stratified flow conditions? Discuss factors such as inhibitor compatibility, dosage requirements, and potential interactions with other pipeline components or fluids.

5 How do the procedures for protecting, monitoring, and evaluating the performance of the LAI inhibitor address the challenges of corrosion control in a low-temperature, stratified flow pipeline? Discuss the importance of continuous monitoring, adjustment of inhibitor dosage, and assessment of corrosion rates in ensuring the inhibitor's effectiveness and the overall integrity of the pipeline.

7.8 Optimizing Corrosion Control in a High Gas Content Oil Pipeline: The Significance of Pigging and the Effectiveness of Laurylamine Imidazoline (LAI) Inhibitor

Introduction: XYZ Company operates a long oil pipeline that faces unique challenges, including high flow rates, high gas content, and the need for effective corrosion control. This case study focuses on the significance of utilizing pigging, the effectiveness of LAI inhibitor, parameters, operation conditions, cathodic protection parameters, relevant codes and standards, and procedures for protecting, monitoring, and evaluating the performance of the inhibitor.

Pipeline description: The oil pipeline spans a significant distance, with a diameter of 24 in. and high gas content. It experiences high flow rates due to the large volume of oil being transported.

Significance of pigging: Pigging is an essential maintenance technique in oil pipelines, serving multiple purposes. It helps remove accumulated debris, sediment, and corrosion products, ensuring the pipeline's optimal flow capacity, integrity, and reducing the risk of corrosion-induced failures.

Effectiveness of laurylamine imidazoline (LAI) inhibitor: The utilization of LAI inhibitor enhances corrosion control in the pipeline, especially in the presence of high gas content. LAI forms a protective film on the pipeline's internal surface, mitigating corrosion and reducing the risk of corrosion-related incidents.

Parameters:

1) **Pipeline diameter**: 24 in. (60.96 cm)
2) **Gas content**: High, measured in percentage (v/v)
3) **Flow rate**: 50,000 barrels per day (bpd) (7.95 m^3/s)

Operation Conditions:

1) **Operating temperature**: 60 °C (140 °F)
2) **Operating pressure**: 1500 psi (10.34 MPa)
3) **Material of construction**: API 5L Grade X65 Carbon Steel

Cathodic Protection Parameters:

1) **Sacrificial anode system**: Zinc or aluminum alloy anodes installed at regular intervals along the pipeline.
2) **Impressed current system**: Mixed metal oxide (MMO) anodes, rectifiers, and reference electrodes.

3) **Current density**: Design current density based on pipeline characteristics, coating condition, and soil resistivity.

Relevant Codes and Standards:

1) **NACE SP0169**: Control of External Corrosion on Underground or Submerged Metallic Piping Systems
2) **API RP 651**: Cathodic Protection of Aboveground Petroleum Storage Tanks
3) **ASME B31.4**: Pipeline Transportation Systems for Liquids and Slurries

Procedure for Protecting and Monitoring the Oil Pipeline:

1) **Pigging strategy**: A comprehensive pigging strategy will be developed that includes regular cleaning, inspection, and corrosion control pigging runs. The frequency and types of pigging required will be determined based on the pipeline's condition, operational parameters, and regulatory requirements.
2) **LAI inhibitor application**: The optimal dosage and injection points for the LAI inhibitor will be determined based on the pipeline's flow rate, gas content, and expected corrosion rate. The inhibitor will be uniformly distributed along the pipeline's internal surface using injection devices and monitoring systems.
3) **Cathodic protection installation and monitoring**: Sacrificial anodes or impressed current anodes will be installed at specified intervals along the pipeline. The performance of the cathodic protection system will be monitored through regular potential measurements, current mapping, and close-interval potential surveys (CIPS).
4) **Pigging operations**: Regular pigging operations will be performed to remove debris, sediment, and corrosion products from the pipeline. Intelligent pigs equipped with inspection tools will be used to assess the internal condition of the pipeline, detect corrosion anomalies, and prioritize maintenance activities.
5) **Corrosion monitoring**: Regular corrosion monitoring will be conducted using techniques such as corrosion coupons, electrical resistance probes, or inline inspection tools. Corrosion data will be analyzed to identify corrosion rates, localized corrosion, and areas requiring targeted corrosion control measures.

Procedure for Evaluating the Performance of the LAI Inhibitor:

1) **Corrosion rate monitoring**: Corrosion rate data will be collected through corrosion coupons, electrical resistance probes, or inline inspection tools. The corrosion rates will be compared with industry standards

and internal performance targets to assess the effectiveness of the LAI inhibitor in mitigating corrosion.

2) **Material analysis**: Pipeline samples will be periodically analyzed to assess the presence of corrosion products, pitting, or other localized forms of corrosion. Techniques like metallography, SEM, and EDS will be utilized to evaluate the LAI inhibitor's performance in high gas content conditions.
3) **Coating integrity assessment**: The external coating will be regularly inspected for any signs of damage or degradation. The coating's integrity and effectiveness in providing an additional barrier against corrosion, especially in high gas content areas, will be assessed.
4) **Environmental monitoring**: Gas content, temperature, pressure, and other relevant parameters will be continuously monitored to evaluate the LAI inhibitor's performance and identify any deviations or anomalies.
5) **Performance evaluation and adjustments**: The collected data will be analyzed and compared against industry standards, regulatory requirements, and established performance targets. Adjustments will be made to the LAI inhibitor dosage, injection points, or corrosion control strategies as needed to optimize corrosion protection and prevent failures.

Questions

1 How does the utilization of pigging contribute to effective corrosion control in an oil pipeline with high gas content? Discuss the specific challenges posed by high gas content and how pigging addresses these challenges to ensure the pipeline's integrity and minimize the risk of corrosion-induced failures.

2 What are the mechanisms by which laurylamine imidazoline (LAI) inhibitor effectively mitigates corrosion in a high gas content pipeline? Discuss the inhibitory properties of LAI inhibitor and its ability to form a protective film, particularly in areas where gas content is high and localized corrosion is more likely to occur.

3 What are the critical considerations in selecting the optimal dosage and injection points for the LAI inhibitor in a pipeline with high gas content? Discuss the factors such as gas content variations, flow dynamics, and material compatibility that impact the effectiveness of the inhibitor and how these parameters are taken into account during the dosage determination process.

4 How does the pipeline's long length impact the corrosion control strategies and the use of LAI inhibitor? Discuss the challenges associated with maintaining inhibitor effectiveness, ensuring uniform distribution over the entire pipeline, and monitoring corrosion rates along the extended length of the pipeline.

5 What are the key factors influencing the performance of the cathodic protection system in a high gas content pipeline? Discuss the considerations related to anode selection, current density, and monitoring techniques, and how these factors are optimized to provide effective cathodic protection and complement the corrosion control efforts of the LAI inhibitor in the presence of high gas content.

References

American Petroleum Institute. (2011). *API RP 571: Damage Mechanisms Affecting Fixed Equipment in the Refining Industry* (2nd ed.). API Publishing Services.

American Petroleum Institute. (2016A). *API 570: Piping Inspection Code: In-Service Inspection, Rating, Repair, and Alteration* (4th ed.). API Publishing Services.

American Petroleum Institute. (2016B). *API RP 574: Inspection Practices for Piping System Components*. API Publishing Services.

American Petroleum Institute. (2016C). *API RP 579: Fitness-for-Service*. API Publishing Services.

American Petroleum Institute. (2018). *API Recommended Practice 5L: Specification for Line Pipe* (50th ed.). API Publishing Services.

American Petroleum Institute. (2018A). *API 653: Tank Inspection, Repair, Alteration, and Reconstruction* (5th ed.). API Publishing Services.

American Petroleum Institute. (2018B). *API RP 970: Corrosion Control Documents*. API Publishing Services.

American Society for Testing and Materials. (2019A). *ASTM E122: Standard Practice for Calculating Sample Size to Estimate, With Specified Precision, the Average for a Characteristic of a Lot or Process*. ASTM International.

American Society for Testing and Materials. (2019B). *ASTM E1225: Standard Test Method for Test Method Validation by Interlaboratory Studies*. ASTM International.

American Society for Testing and Materials. (2019C). *ASTM E178: Standard Practice for Dealing With Outlying Observations*. ASTM International.

American Society for Testing and Materials. (2019D). *ASTM G1: Standard Practice for Preparing, Cleaning, and Evaluating Corrosion Test Specimens*. ASTM International.

Imidazoline Inhibitors for Corrosion Protection of Oil Pipeline Steels: Experimental Laboratory Evaluation and Case Studies, First Edition. Faysal Fayez Eliyan.
© 2024 John Wiley & Sons, Inc. Published 2024 by John Wiley & Sons, Inc.

Amin, M. A., Ahmed, M. A., Arida, H. A., Arslan, T., Saracoglu, M., & Kandemirli, F. (2011). Monitoring corrosion and corrosion control of iron in HCl by non-ionic surfactants of the TRITON-X series–Part II. Temperature effect, activation energies and thermodynamics of adsorption. *Corrosion Science*, 53(2), 540–548.

Arukalam, I. O., Madufor, I. C., Ogbobe, O., & Oguzie, E. E. (2015). Inhibition of mild steel corrosion in sulfuric acid medium by hydroxyethyl cellulose. *Chemical Engineering Communications*, 202(1), 112–122.

Babarao, R., & Jiang, J. (2008). Diffusion and separation of CO_2 and CH_4 in silicalite, C168 schwarzite, and IRMOF-1: a comparative study from molecular dynamics simulation. *Langmuir*, 24, 5474–5484.

Babić-Samardžija, K., Lupu, C., Hackerman, N., Barron, A. R., & Luttge, A. (2005). Inhibitive properties and surface morphology of a group of heterocyclic diazoles as inhibitors for acidic iron corrosion. *Langmuir*, 21(26), 12187–12196.

Bailey, S., Tan, Y. J., & Kinsella, B. (1997). Use of electrochemical impedance spectroscopy for study of CO_2 corrosion prevention by batch treatment inhibitor films. *British Corrosion Journal*, 32(1), 49–56.

Belarbi, Z., Brown, B., Singer, M., & Srdjan, N. (2017, March). Study of adsorption of corrosion inhibitor 1-(2-aminoethyl)-2-oleyl-2-imidazolinium chloride on carbon steel under CO2 environment by using in-situ AFM measurements. In *Corrosion 2017*. OnePetro.

Bentiss, F., Jama, C., Mernari, B., El Attari, H., El Attari, H., El Kadi, L., Lebrini, M., & Lagrenée, M. (2009). Corrosion control of mild steel using 3,5-bis (4-methoxyphenyl)-4-amino-1,2,4-triazole in normal hydrochloric acid medium. *Corrosion Science*, 51(8), 1628–1635.

Bentiss, F., Lebrini, M., & Lagrenée, M. (2005). Thermodynamic characterization of metal dissolution and inhibitor adsorption processes in mild steel/2,5-bis (n-thienyl)-1,3,4-thiadiazoles/hydrochloric acid system. *Corrosion Science*, 47(12), 2915–2931.

Bentiss, F., Traisnel, M., Chaibi, N., Mernari, B., Vezin, H., & Lagrenée, M. (2002). 2,5-Bis (n-methoxyphenyl)-1,3,4-oxadiazoles used as corrosion inhibitors in acidic media: Correlation between inhibition efficiency and chemical structure. *Corrosion Science*, 44(10), 2271–2289.

Binks, B., & Furlong, D. (1999). *Modern Characterization Methods of Surfactant Systems (Surfactant Science)*, CRC Press, Boca Raton.

Bockris, J. O'M., Drazic, D., & Despic, A. R. (1961). The electrode kinetics of the deposition and dissolution of iron, *Electrochimica Acta*, 4(2–4), 325–361.

Branzoi, V., Branzoi, F., & Baibarac, M. (2000). The inhibition of the corrosion of Armco iron in HCl solutions in the presence of surfactants of the type of N-alkyl quaternary ammonium salts. *Materials Chemistry and Physics*, 65(3), 288–297.

Cao, C. (1996). On electrochemical techniques for interface inhibitor research. *Corrosion Science*, 38(12), 2073–2082.

Chai, C. W., Lu, M. X., & Zhang, Y. F. (2005). The Proceedings of the 3rd Corrosion and Protection Conference on Processing of ShanDong Petroleum Society (p. 126).

Chang, C. J., Li, S. S., & Ko, C. M. (1995). Catalytic wet oxidations of phenol- and p-chlorophenol-contaminated waters. *Journal of Chemical Technology & Biotechnology*, 64(3), 245–252.

Chen, G. H., Zhou, T., Shen, S. Y., Ouyang, J., Zhu, Z. Y., Wang, F., ... & Zhao, J. M. (2019). Inhibition Behavior of Two Imidazoline-Based Inhibitors in Single Liquid Phase Flow. In *Proceedings of the International Field Exploration and Development Conference 2017* (pp. 223–233). Springer Singapore.

Chen, G., Li, L., Ouyang, J., Zhu, Z., Wang, F., Wang, Y., Xue, J., & Zhao, J. (2015). The synergistic inhibition effect between imidazoline and the oxide film on L360 steel in CO_2/H_2S systems. *Anti-Corrosion Methods and Materials*, 62(3), 143–148.

Conde, A., & De Damborenea, J. J. (2002). Electrochemical impedance spectroscopy for studying the degradation of enamel coatings. *Corrosion Science*, 44(7), 1555–1567.

Cruz, J., Martınez, R., Genesca, J., & Garcıa-Ochoa, E. (2004). Experimental and theoretical study of 1-(2-ethylamino)-2-methylimidazoline as an inhibitor of carbon steel corrosion in acid media. *Journal of Electroanalytical Chemistry*, 566(1), 111–121.

de Souza, F. S., & Spinelli, A. (2009). Caffeic acid as a green corrosion inhibitor for mild steel. *Corrosion Science*, 51(3), 642–649.

Deng, S., Li, X., & Xie, X. (2014). Hydroxymethyl urea and 1,3-bis (hydroxymethyl) urea as corrosion inhibitors for steel in HCl solution. *Corrosion Science*, 80, 276–289.

Denpo, K., & Ogawa, H. (1993). Fluid flow effects on CO_2 corrosion resistance of oil well materials. *Corrosion*, 49(6), 442–449.

Desimone, M. P., Grundmeier, G., Gordillo, G., & Simison, S. N. (2011). Amphiphilic amido-amine as an effective corrosion inhibitor for mild steel exposed to CO_2 saturated solution: Polarization, EIS and PM-IRRAS studies. *Electrochimica Acta*, 56(8), 2990–2998.

Dewar, M. J., & Thiel, W. (1977). Ground states of molecules. 38. The MNDO method. Approximations and parameters. *Journal of the American Chemical Society*, 99(15), 4899–4907.

Diaz, E. F., Gonzalez-Rodriguez, J. G., Martinez-Villafañe, A., & Gaona-Tiburcio, C. (2010). H_2S corrosion inhibition of an ultra high strength pipeline by carboxyethyl-imidazoline. *Journal of Applied Electrochemistry*, 40(9), 1633–1640.

Ding, Y., Brown, B., Young, D., & Singer, M. (2018, April). Effectiveness of an imidazoline-type inhibitor against CO_2 corrosion of mild steel at elevated temperatures (120 C-150 C). In *Corrosion 2018*. OnePetro.

Doležal, D., Alar, V., & Borko, T. (2013). Inhibition of carbon steel corrosion by imidazoline in carbon dioxide saturated oilfield brine solution. *Materialwissenschaft und Werkstofftechnik*, 44(7), 626–632.

Edwards, A., Osborne, C., Webster, S., Klenerman, D., Joseph, M., Ostovar, P., & Doyle, M. (1994). Mechanistic studies of the corrosion inhibitor oleic imidazoline. *Corrosion Science*, 36(2), 315–325.

ElBelghiti, M., Karzazi, Y., Dafali, A., Hammouti, B., Bentiss, F., Obot, I., & Ebenso, E. E. (2016). Experimental, quantum chemical and Monte Carlo simulation studies of 3,5-disubstituted-4-amino-1,2,4-triazoles as corrosion inhibitors on mild steel in acidic medium. *Journal of Molecular Liquids*, 218, 281–293.

Feng, L. J., Yang, H. Y., & Wang, F. H. (2012). Effect of an imidazoline derivative on the protection performance of oxide film formed on carbon steel in saturated $Ca(OH)_2$ solution. *International Journal of Electrochemical Science*, 7, 4064–4077.

Finšgar, M., Lesar, A., Kokalj, A., & Milošev, I. (2008). A comparative electrochemical and quantum chemical calculation study of BTAH and BTAOH as copper corrosion inhibitors in near neutral chloride solution. *Electrochimica Acta*, 53(28), 8287–8297.

Geerlings, P., & De Proft, F. (2002). Chemical reactivity as described by quantum chemical methods. *International Journal of Molecular Sciences*, 3(4), 276–309.

Hansen, J., & McDonald, I. (2013). *Theory of Simple Liquids: With Applications to Soft Matter*. Academic Press.

Heuer, J. K., & Stubbins, J. F. (1999). An XPS characterization of $FeCO_3$ films from CO_2 corrosion. *Corrosion Science*, 41(7), 1231–1243.

Hassan, H. H., Essam Abdelghani, & Mohammed A. Amin. (2007). Inhibition of mild steel corrosion in hydrochloric acid solution by triazole derivatives: Part I. Polarization and EIS studies. *Electrochimica Acta*, 52(22), 6359–6366.

Heydari, M., & Javidi, M. (2012). Corrosion inhibition and adsorption behaviour of an amido-imidazoline derivative on API 5L X52 steel in CO_2-saturated solution and synergistic effect of iodide ions. *Corrosion Science*, 61, 148–155.

Hladky, K., & Dawson, J. L. (1982). The measurement of corrosion using electrochemical 1f noise. *Corrosion Science*, 22(3), 231–237.

Hong, T., Sun, Y. H., & Jepson, W. P. (2002). Study on corrosion inhibitor in large pipelines under multiphase flow using EIS. *Corrosion Science*, 44(1), 101–112.

International Organization for Standardization. (2017). *ISO 12944-1:2017 – Paints and Varnishes: Corrosion Protection of Steel Structures by Protective Paint Systems – Part 1: General Introduction*. International Organization for Standardization.

International Organization for Standardization. (2019A). *ISO 3183: Petroleum and Natural Gas Industries – Steel Pipe for Pipeline Transportation Systems*. ISO.

International Organization for Standardization. (2019B). *ISO 5725: Accuracy (Trueness and Precision) of Measurement Methods and Results*. ISO.

Javidi, M., Chamanfar, R., & Bekhrad, S. (2019). Investigation on the efficiency of corrosion inhibitor in CO2 corrosion of carbon steel in the presence of iron carbonate scale. *Journal of Natural Gas Science and Engineering*, 61, 197–205.

Jevremović, I., Singer, M., Nešić, S., & Mišković-Stanković, V. (2016). Electrochemistry of carbon dioxide corrosion mitigation using tall oil diethylenetriamine imidazoline as corrosion inhibitor for mild steel. *Materials and Corrosion*, 67(7), 756–768.

Jiang, J., & Sandler, S. I. (2005C). Separation of CO_2 and N_2 by adsorption in C168 schwarzite: A combination of quantum mechanics and molecular simulation study. *Journal of the American Chemical Society*, 127(34), 11989–11997.

Jiang, X., Zheng, Y. G., & Ke, W. (2005A). Effect of flow velocity and entrained sand on inhibition performances of two inhibitors for CO_2 corrosion of N80 steel in 3% NaCl solution. *Corrosion Science*, 47(11), 2636–2658.

Jiang, X., Zheng, Y. G., & Ke, W. (2005B). Corrosion inhibitor performances for carbon dioxide corrosion of N80 steel under static and flowing conditions. *Corrosion*, 61(4), 326–334.

Jiang, Y. Y., Chen, Y., Ye, Z. Y., Chen, H., Zhang, Z., Zhang, J. Q., & Cao, C. N. (2013). Inhibition of iron corrosion in HCl solutions by N-[2-[(2-aminoethyl) amino] Ethyl]-9-octadecenamide. *Corrosion*, 69(7), 672–680.

Jing-mao, Z. H. A. O., He-xia, L. I. U., Wei, D. I., & Yu, Z. U. O. (2004). The inhibition synergistic effect between imidazoline derivative and thiourea. *Journal of Electrochemistry*, 10(4), 440.

Jovancicevic, V., Ramachandran, S., & Prince, P. (1998). Inhibition of carbon dioxide corrosion of mild steel by imidazolines and their precursors. *Corrosion*, 55(5), 449–455.

Kang, E. T., Neoh, K. G., & Tan, K. L. (1992). The intrinsic redox states in polypyrrole and polyaniline: a comparative study by XPS. Surf. *Interface Anal.* 19, 33–37.

Keddam, M., Mottos, O. R., & Takenouti, H. (1981). Reaction model for iron dissolution studied by electrode impedance: I. Experimental results and reaction model. *Journal of the Electrochemical Society*, 128(2), 257.

Kern, P., & Landolt, D. (2002). Adsorption of a bromine labeled carboxylic acid corrosion inhibitor on iron measured with EQCM, EIS and XPS. *Corrosion Science*, 44(8), 1809–1824.

Khavasfar, A., Moayed, M. H., & Attar, M. M. (2007). An investigation on the performance of an imidazoline based commercial corrosion inhibitor on CO_2 corrosion of gas-well tubing steel by EIS technique. *Iranian Journal of Materials Science and Engineering*, 4, 1–8.

Kovačević, N., & Kokalj, A. (2011). Analysis of molecular electronic structure of imidazole-and benzimidazole-based inhibitors: A simple recipe for qualitative estimation of chemical hardness. *Corrosion Science*, 53(3), 909–921.

Lgaz, H., Salghi, R., Bhat, K. S., Chaouiki, A., & Jodeh, S. (2017). Correlated experimental and theoretical study on inhibition behavior of novel quinoline derivatives for the corrosion of mild steel in hydrochloric acid solution. *Journal of Molecular Liquids*, 244, 154–168.

Li, S., & Purdy, W. C. (1992). Cyclodextrins and their applications in analytical chemistry. *Chemical Reviews*, 92(6), 1457–1470. https://doi.org/10.1021/cr00014a009.

Li, W., & Li, D. Y. (2006). Influence of surface morphology on corrosion and electronic behavior. *Acta Materialia*, 54(2), 445–452.

Li, Y. Z., Xu, N., Guo, X. P., & Zhang, G. A. (2017). Inhibition effect of imidazoline inhibitor on the crevice corrosion of N80 carbon steel in the CO_2-saturated NaCl solution containing acetic acid. *Corrosion Science*, 126, 127–141.

Liu, D., Qiu, Y. B., Tomoe, Y., Bando, K., & Guo, X. P. (2011). Interaction of inhibitors with corrosion scale formed on N80 steel in CO_2-saturated NaCl solution. *Materials and Corrosion*, 62(12), 1153–1158.

Liu, F. G., Du, M., Zhang, J., & Qiu, M. (2009B). Electrochemical behavior of Q235 steel in saltwater saturated with carbon dioxide based on new imidazoline derivative inhibitor. *Corrosion Science*, 51(1), 102–109.

Liu, L., Zhang, Y., Yin, X., & Qian, J. (2012). Anti-corrosion properties of bis (benzimidazole) derivatives for N80 steel in H_2S solution. *Surface and Interface Analysis*, 45(6), 949–954.

Liu, X., Okafor, P. C., & Zheng, Y. G. (2009A). The inhibition of CO_2 corrosion of N80 mild steel in single liquid phase and liquid/particle two-phase flow by aminoethyl imidazoline derivatives. *Corrosion Science*, 51(4), 744–751.

Lopez, D. A., Schreiner, W. H., De Sanchez, S. R., & Simison, S. N. (2004). The influence of inhibitors molecular structure and steel microstructure on corrosion layers in CO_2 corrosion: An XPS and SEM characterization. *Applied Surface Science*, 236(1–4), 77–97.

Lopez, D. A., Simison, S. N., & De Sanchez, S. R. (2005). Inhibitors performance in CO_2 corrosion: EIS studies on the interaction between their molecular structure and steel microstructure. *Corrosion Science*, 47(3), 735–755.

Lucio-Garcia, M. A., Gonzalez-Rodriguez, J. G., Martinez-Villafañe, A., Dominguez-Patiño, G., Neri-Flores, M. A., & Chacon-Nava, J. G. (2010). A study of hydroxyethyl imidazoline as H_2S corrosion inhibitor using electrochemical noise and electrochemical impedance spectroscopy. *Journal of Applied Electrochemistry*, 40, 393–399.

Materials Studio. (2013). *Revision 6.0.* Accelrys: San Diego.

Matthews, C. J., Leese, T. A., Clegg, W., Elsegood, M. R., Horsburgh, L., & Lockhart, J. C. (1996). A route to bis (benzimidazole) ligands with built-in asymmetry: Potential models of protein binding sites having histidines of different basicity. *Inorganic Chemistry*, 35(26), 7563–7571.

Migahed, M., El-Rabiei, M. M., Nady, H., Gomaa, H., & Zaki, E. G. (2017). Corrosion inhibition behavior of synthesized imidazolium ionic liquids for carbon steel in deep oil wells formation water. *Journal of Bio- and Tribo-Corrosion*, 3(2). https://doi.org/10.1007/s40735-017-0080-5.

Moiseeva, L. S., & Kuznetsov, Y. I. (1997). Inhibition of carbon-dioxide corrosion of equipment of the petroleum and gas industry. *Chemical and Petroleum Engineering*, 33(1), 81–87.

Muralidharan, S., Quraishi, M. A., & Iyer, S. V. K. (1995). The effect of molecular structure on hydrogen permeation and the corrosion inhibition of mild steel in acidic solutions. *Corrosion Science*, 37(11), 1739–1750.

Musa, A. Y., Kadhum, A. A. H., Mohamad, A. B., Rahoma, A. A. B., & Mesmari, H. (2010). Electrochemical and quantum chemical calculations on 4,4-dimethyloxazolidine-2-thione as inhibitor for mild steel corrosion in hydrochloric acid. *Journal of Molecular Structure*, 969(1–3), 233–237.

NACE ASTM (2012). Standard guide for laboratory immersion corrosion testing of metals. *ASTM Int*, 1–9.

NACE International. (2016A). *NACE SP0198: Control of Corrosion Under Thermal Insulation and Fireproofing Materials: A Systems Approach.* NACE International.

NACE International. (2012A). *NACE RP0490: Holiday Detection of Fusion-bonded Epoxy External Pipeline Coatings of 250 to 760 μm (10 to 30 mils).* NACE International.

NACE International. (2012B). *NACE RP0102: In-line Inspection of Pipelines.* NACE International.

NACE International. (2015). *NACE SP0169: Control of External Corrosion on Underground or Submerged Metallic Piping Systems.* NACE International.

NACE International. (2016). *NACE TM0169: Laboratory Corrosion Testing of Metals.* NACE International.

NACE International. (2016B). *NACE TM0284: Evaluation of Pipeline and Pressure Vessel Steels for Resistance to Hydrogen-induced Cracking.* NACE International.

NACE International. (2017). *NACE Standard RP0775: Preparation, Installation, Analysis, and Interpretation of Corrosion Coupons in Oilfield Operations.* NACE International.

NACE International. (2017A). *NACE RP0204: Stress Corrosion Cracking (SCC) Direct Assessment Methodology*. NACE International.

NACE International. (2017B). *NACE SP0285: Internal Corrosion Direct Assessment Methodology for Pipelines Carrying Normally Dry Natural Gas.* NACE International.

NACE Standard RP0775-99 Item No. 21017 (1999). Preparation, Installation, Analysis, and Interpretation of Corrosion Coupons in Oilfield Operations.

Nesic, S., Postlethwaite, J., & Olsen, S. (1996). An electrochemical model for prediction of corrosion of mild steel in aqueous carbon dioxide solutions. *Corrosion*, 52(4), 280–294.

Ning, S., Shi, M., Liu, F., Yu, L., Li, Y., Zhang, R., & Cheng, Z. (1990). The relationship of corrosion inhibitor efficiency on steel in acids with electron density and the energy of frontier orbital of imidazoline derivatives. *Journal of Chinese Society for Corrosion and Protection* 10, 383–390.

Ning, J., Zheng, Y., Young, D., Brown, B., & Nesic, S. (2013, March) A thermodynamic study of hydrogen sulfide corrosion of mild steel. In *Corrosion 2013*. OnePetro.

Obot, I. B. (2009). Synergistic effect of nizoral and iodide ions on the corrosion inhibition of mild steel in sulphuric acid solution. *Portugaliae Electrochimica Acta*, 27(5), 539–553.

Oguzie, E. E., Li, Y., & Wang, F. H. (2007). Corrosion inhibition and adsorption behavior of methionine on mild steel in sulfuric acid and synergistic effect of iodide ion. *Journal of Colloid and Interface Science*, 310(1), 90–98.

Okafor, P. C., Liu, C., Zhu, Y. S., & Zheng, Y. H. (2011). Corrosion and corrosion inhibition behavior of N80 and P110 carbon steels in CO_2-saturated simulated formation water by rosin amide imidazoline. *Industrial & Engineering Chemistry Research*, 50(12), 7273–7281. https://doi.org/10.1021/ie1024112.

Okafor, P. C., Liu, X., & Zheng, Y. G. (2009). Corrosion inhibition of mild steel by ethylamino imidazoline derivative in CO_2-saturated solution. *Corrosion Science*, 51(4), 761–768.

Olivares-Xometl, O., Likhanova, N. V., Domínguez-Aguilar, M. A., Hallen, J. M., Zamudio, L. S., & Arce. E. (2006). Surface analysis of inhibitor films formed by imidazolines and amides on mild steel in an acidic environment. *Applied Surface Science*, 252(6), 2139–2152.

Ortega-Sotelo, D. M., Gonzalez-Rodriguez, J. G., Neri-Flores, M. A., Casales, M., Martinez, L., & Martinez-Villafañe, A. (2011). CO_2 corrosion inhibition of X-70 pipeline steel by carboxyamido imidazoline. *Journal of Solid State Electrochemistry*, 15, 1997–2004.

Ortega-Toledo, D. M., Gonzalez-Rodriguez, J. G., Casales, M., Neri-Florez, M. A., & Martinez-Villafane, A. (2010). The CO_2 corrosion inhibition of a high strength pipeline steel by hydroxyethyl imidazoline. *Materials Chemistry and Physics*, 122(2–3), 485–490.

Pearson, R. G. (1990). Hard and soft acids and bases – the evolution of a chemical concept. *Coordination Chemistry Reviews*, 100, 403–425.

Popova, A., Sokolova, E., Raicheva, S., & Christov, M. (2003). AC and DC study of the temperature effect on mild steel corrosion in acid media in the presence of benzimidazole derivatives. *Corrosion Science*, 45(1), 33–58.

Porcayo-Calderon, J., De La Escalera, L. M., Canto, J., & Casales-Diaz, M. (2015). Imidazoline derivatives based on coffee oil as CO_2 corrosion inhibitor. *International Journal of Electrochemical Science*, 10, 3160–3176.

Raicheva, S. N., Aleksiev, B. V., & Sokolova, E. I. (1993). The effect of the chemical structure of some nitrogen-and sulphur-containing organic compounds on their corrosion inhibiting action. *Corrosion Science*, 34(2), 343–350.

Ramachandran, S., Tsai, B. L., Blanco, M., Chen, H., Tang, Y., & Goddard, W. A. (1996). Self-assembled monolayer mechanism for corrosion inhibition of iron by imidazolines. *Langmuir*, 12(26), 6419–6428.

Ravi, S. K., Sun, W., Nandakumar, D. K., Zhang, Y., & Tan, S. C. (2018). Optical manipulation of work function contrasts on metal thin films. *Science Advances*, 4(3), eaao6050.

Refaey, S. A. M., Taha, F., & Abd El-Malak, A. M. (2004). Inhibition of stainless steel pitting corrosion in acidic medium by 2-mercaptobenzoxazole. *Applied Surface Science*, 236(1–4), 175–185.

Rivera-Grau, L. M., Casales, M., Regla, I., Ortega-Toledo, D. M., Ascencio-Gutierrez, J. A., Porcayo-Calderon, J., & Martinez-Gomez, L. (2013). Effect of organic corrosion inhibitors on the corrosion performance of 1018 carbon steel in 3% NaCl solution. *International Journal of Electrochemical Science*, 8(2), 2491–2503.

Rivera-Grau, L. M., Gonzalez-Rodriguez, J. G., & Martinez, L. (2016). Effect of hydroxyethil imidazoline and Ag nanoparticles on the CO_2 corrosion of carbon steel. *International Journal of Electrochemical Science*, 11(8), 94.

Shamsa, A., Barker, R., Hua, Y., Barmatov, E., Hughes, T. L., & Neville, A. (2020). Performance evaluation of an imidazoline corrosion inhibitor in a CO_2-saturated environment with emphasis on localised corrosion. *Corrosion Science*, 176, 108916.

Singh, A., Ansari, K. R., Chauhan, D. S., Quraishi, M. A., & Kaya, S. (2020). Anti-corrosion investigation of pyrimidine derivatives as green and sustainable corrosion inhibitor for N80 steel in highly corrosive environment: Experimental and AFM/XPS study. *Sustainable Chemistry and Pharmacy*, 16, 100257.

Singh, A., Ansari, K. R., Lin, Y., Quraishi, M. A., Lgaz, H., & Chung, I. M. (2019). Corrosion inhibition performance of imidazolidine derivatives for J55 pipeline steel in acidic oilfield formation water: Electrochemical, surface and theoretical studies. *Journal of the Taiwan Institute of Chemical Engineers*, 95, 341–356.

Singh, I. (1993). Inhibition of steel corrosion by thiourea derivatives. *Corrosion*, 49(6), 473–478.

Solomon, M. M., Umoren, S. A., Quraishi, M. A., Tripathy, D. B., & Abai, E. J. (2020). Effect of akyl chain length, flow, and temperature on the corrosion inhibition of carbon steel in a simulated acidizing environment by an imidazoline-based inhibitor. *Journal of Petroleum Science and Engineering*, 187, 106801.

Sukul, D., Pal, A., Saha, S. K., Satpati, S., Adhikari, U., & Banerjee, P. (2018). Newly synthesized quercetin derivatives as corrosion inhibitors for mild steel in 1 M HCl: Combined experimental and theoretical investigation. *Physical Chemistry Chemical Physics*, 20(9), 6562–6574.

Talbot D., *Corrosion Science and Technology*, CRC Press, London 1997.

Tan, Y. J., Bailey, S., & Kinsella, B. (1996). An investigation of the formation and destruction of corrosion inhibitor films using electrochemical impedance spectroscopy (EIS). *Corrosion Science*, 38(9), 1545–1561.

Tantawy, A. H., Soliman, K. A., & Abd El-Lateef, H. M. (2020). Novel synthesized cationic surfactants based on natural piper nigrum as sustainable-green inhibitors for steel pipeline corrosion in CO_2-3.5% NaCl: DFT, Monte Carlo simulations and experimental approaches. *Journal of Cleaner Production*, 250, 119510.

Tao, Z., He, W., Wang, S., Zhang, S., & Zhou, G. (2012). A study of differential polarization curves and thermodynamic properties for mild steel in acidic solution with nitrophenyltriazole derivative. *Corrosion Science*, 60, 205–213.

Temesghen, W., & Sherwood, P. (2002). Analytical utility of valence band X-ray photoelectron spectroscopy of iron and its oxides, with spectral interpretation by cluster and band structure calculations. *Analytical and Bioanalytical Chemistry*, 373, 601–608.

Tiu, B. D. B., & Advincula, R. C. (2015). Polymeric corrosion inhibitors for the oil and gas industry: Design principles and mechanism. *Reactive and Functional Polymers*, 95, 25–45.

Ueda, M., & Ikeda, A. (1996, March). Effect of microstructure and Cr content in steel on CO_2 corrosion. In *Corrosion 96*. OnePetro.

Umoren, S. A., Obot, I. B., Madhankumar, A., & Gasem, Z. M. (2015). Performance evaluation of pectin as ecofriendly corrosion inhibitor for X60 pipeline steel in acid medium: Experimental and theoretical approaches. *Carbohydrate Polymers*, 124, 280–291.

Usman, B. J., & Ali, S. A. (2018). Carbon dioxide corrosion inhibitors: A review. *Arabian Journal for Science and Engineering*, 43(1), 1–22. https://doi.org/10.1007/s13369-017-2949-5.

Verma, C., Lgaz, H., Verma, D. K., Ebenso, E. E., Bahadur, I., & Quraishi, M. A. (2018). Molecular dynamics and Monte Carlo simulations as powerful tools for study of interfacial adsorption behavior of corrosion inhibitors in aqueous phase: A review. *Journal of Molecular Liquids*, 260, 99–120.

Videm, K., & Dugstad, A. (1989). Corrosion of carbon steel in an aqueous carbon dioxide environment Part 1. *Materials Performance (United States)*, 28(3).

Vračar, L. M., & Dražić, D. M. (2002). Adsorption and corrosion inhibitive properties of some organic molecules on iron electrode in sulfuric acid. *Corrosion Science*, 44(8), 1669–1680.

Wang, B., Du, M., & Zhang, J. (2009). Study of the inhibition mechanism of imidazoline derivative inhibitor on CO_2 corrosion for Q235 steel. In *Advanced Materials Research* (Vol. 79, pp. 981–984). Trans Tech Publications Ltd.

Wang, B., Du, M., Zhang, J., & Gao, C. J. (2011). Electrochemical and surface analysis studies on corrosion inhibition of Q235 steel by imidazoline derivative against CO_2 corrosion. *Corrosion Science*, 53(1), 353–361.

Wang, D., Li, S., Ying, Y., Wang, M., Xiao, H., & Chen, Z. (1999). Theoretical and experimental studies of structure and inhibition efficiency of imidazoline derivatives. *Corrosion Science*, 41(10), 1911–1919.

Wang, H. (2005, April). Application of electrochemical noise technique in multiphase flow. In *Corrosion 2005*. OnePetro.

Wang, L., Zhang, S. W., Guo, Q., Zheng, H., Lu, D. M., Peng, L., & Xion, J. (2015). Corrosion inhibition of mild steel by methyl violet and bromide ion in sulfuric acid solution. *Materials and Corrosion*, 66(6), 594–602.

Wang, X. Z., Yang, Z. G., Yin, Z. F., Gao, Z. Q., Li, J. D., & Wang, K. (2014). Corrosion investigation of J55 steel under simulated enhanced oil recovery conditions using CO_2 flooding. *Corrosion Engineering, Science and Technology*, 49(4), 275–282.

Wang, Z. Z., Li, Y. Y., & Zhang, G. A. (2019). Inhibitive effects of inhibitors on the galvanic corrosion between N80 carbon steel and 13Cr stainless steel under dynamic supercritical CO_2 conditions. *Corrosion Science*, 146, 121–133.

Wei, L., Pang, X., & Gao, K. (2016). Corrosion of low alloy steel and stainless steel in supercritical $CO_2/H_2O/H_2S$ systems. *Corrosion Science*, 111, 637–648.

Wu, R. , Qiu, X. , Shi, Y., & Deng, M. (2017). Molecular dynamics simulation of the atomistic monolayer structures of N-acyl amino acid-based surfactants. *Mol Simul*, 43, 491–501.

Xiang, Y., Long, Z., Li, C., Huang, H., & He, X. (2017). Inhibition of N80 steel corrosion in impure supercritical CO_2 and CO_2-saturated aqueous phases by using imino inhibitors. *International Journal of Greenhouse Gas Control*, 63, 141–149.

Xie, S. W., Liu, Z., Han, G. C., Li, W., Liu, J., & Chen, Z. (2015). Molecular dynamics simulation of inhibition mechanism of 3,5-dibromo salicylaldehyde Schiff's base. *Computational and Theoretical Chemistry*, 1063, 50–62.

Xiong, L., He, Z., Han, S., Tang, J., Wu, Y., & Zeng, X. (2016). Tribological properties study of N-containing heterocyclic imidazoline derivatives as lubricant additives in water-glycol. *Tribology International*, 104, 98–108.

Xu, Y., Li, H., Li, S., Guan, X., & Lan, C. (2016). 3-D modelling and statistical properties of surface pits of corroded wire based on image processing technique. *Corrosion Science*, 111, 275–287.

Yan, X., Zou, C., & Qin, Y. (2014). A new sight of water-soluble polyacrylamide modified by β-cyclodextrin as corrosion inhibitor for X70 steel. *Starch-Starke*, 66(11–12), 968–975. 10.1002/star.201400063

Yang, S., Richter, S., Robbins, W., & Nesic, S. (2012, March). Evaluation of the protectiveness of a paraffin layer in CO_2 corrosion of mild steel. In *Corrosion 2012*. OnePetro.

Yoo, S. H., Kim, Y. W., Chung, K., Baik, S. Y., & Kim, J. S. (2012). Synthesis and corrosion inhibition behavior of imidazoline derivatives based on vegetable oil. *Corrosion Science*, 59, 42–54.

Yurt, A., & Aykın, Ö. (2011). Diphenolic Schiff bases as corrosion inhibitors for aluminium in 0.1 M HCl: Potentiodynamic polarisation and EQCM investigations. *Corrosion Science*, 53(11), 3725–3732.

Zavala Olivares, G., Hernández Gayosso, M. J., & Mora Mendoza, J. L. (2007). Corrosion inhibitors performance for mild steel in CO_2 containing solutions. *Materials and Corrosion*, 58(6), 427–437.

Zhang, G. A., Zeng, Y., Guo, X. P., Jiang, F., Shi, D. Y., & Chen, Z. Y. (2012). Electrochemical corrosion behavior of carbon steel under dynamic high pressure H_2S/CO_2 environment. *Corrosion Science*, 65, 37–47.

Zhang, G., Chen, C., Lu, M., Chai, C., & Wu, Y. (2007). Evaluation of inhibition efficiency of an imidazoline derivative in CO_2-containing aqueous solution. *Materials Chemistry and Physics*, 105(2–3), 331–340.

Zhang, H. H., Gao, K., Yan, L., & Pang, X. (2017). Inhibition of the corrosion of X70 and Q235 steel in CO_2-saturated brine by imidazoline-based inhibitor. *Journal of Electroanalytical Chemistry*, 791, 83–94.

Zhang, H. H., Pang, X., & Gao, K. (2018). Localized CO_2 corrosion of carbon steel with different microstructures in brine solutions with an imidazoline-based inhibitor. *Applied Surface Science*, 442, 446–460.

Zhang, H. H., Pang, X., & Gao, K. (2019). Effect of surface roughness on the performance of thioureido imidozaline inhibitor in CO_2-saturated brine. *Corrosion Science*, 157, 189–204.

Zhang, H. H., Pang, X., Zhou, M., Liu, C., Wei, L., & Gao, K. (2015). The behavior of pre-corrosion effect on the performance of imidazoline-based inhibitor in 3 wt.% NaCl solution saturated with CO_2. *Applied Surface Science*, 356, 63–72.

Zhang, X. (2009). Absorption and adsorption behavior of imidazoline amide on iron surface in CO_2 solution. *Journal of Chinese Society for Corrosion and Protection*, 20(6), 344–348.

Zhang, X., Wang, F., He, Y., & Du, Y. (2001). Study of the inhibition mechanism of imidazoline amide on CO_2 corrosion of Armco iron. *Corrosion Science*, 43(8), 1417–1431.

Zhao, J., & Chen, G. (2012). The synergistic inhibition effect of oleic-based imidazoline and sodium benzoate on mild steel corrosion in a CO_2-saturated brine solution. *Electrochimica Acta*, 69, 247–255.

Zhao, J., Gu, F., Wang, D., & Yu, M. (2018). Effects of thioureido imidazoline on the passivation and pitting corrosion of N80 steel in CO_2-saturated NaCl-NaNO2 solution. *International Journal of Electrochemical Science*, 13, 2676–2687.

Zou, C., Zhao, P., Lei, Y., Ye, H., Yao, Y., Chen, M., & Wang, T. (2011). Preparation and performance of a novel water-soluble cationic polymer containing β-cyclodextrin. *Chemical Engineering & Technology*, 34(11), 1820–1826.

Index

Imidazoline Inhibitors for Corrosion Protection of Oil Pipeline Steels: Experimental Laboratory Evaluation and Case Studies, First Edition. Faysal Fayez Eliyan.
© 2024 John Wiley & Sons, Inc. Published 2024 by John Wiley & Sons, Inc.

d

e

j

k

l

q

r

t

u

Printed and bound by CPI Group (UK) Ltd, Croydon, CR0 4YY
17/09/2026
14964317-0001